Parasitology Research Monographs

Volume 5

Series Editor:
Heinz Mehlhorn
Department of Zoomorphology
Cell Biology and Parasitology
Heinrich Heine University
Universitätsstrasse 1
40225 Düsseldorf
Germany

For further volumes:
http://www.springer.com/series/8816

Sven Klimpel • Heinz Mehlhorn
Editors

Bats (Chiroptera) as Vectors of Diseases and Parasites

Facts and Myths

Springer

Editors
Sven Klimpel
Biodiv. and Climate Research Centre
Goethe-University
Frankfurt
Germany

Heinz Mehlhorn
Department of Zoomorphology
Heinrich Heine University
Düsseldorf
Germany

ISSN 2192-3671 ISSN 2192-368X (electronic)
ISBN 978-3-642-39332-7 ISBN 978-3-642-39333-4 (eBook)
DOI 10.1007/978-3-642-39333-4
Springer Heidelberg New York Dordrecht London

Library of Congress Control Number: 2013951330

Printed on acid-free paper

Springer is part of Springer Science+Business Media (www.springer.com)

Preface

Bats (Chiroptera in zoological terms) are unique in the Kingdom of animals: they fly obviously like birds—but are not birds, since the females give birth to their descendents and let them suck milk. In daytime they sleep/stay in hiding places. All bat species become active only in the dusk time and keep active until dawn. Some species start their quick flattering flights in order to catch their food consisting of flying insects, while other species fly at night from their dayover sleeping places at the top of trees to fruit plantations, where they cut off pieces from mangos, pears, apples, etc. A few other species in South and Central America even attack mammals, scratch their skin, and lick the leaking blood. This strict nocturnal activity led to the fact that many humans never had a chance to snap a close look at the body and features of bats. Thus, it is understandable that their silent flight while scurrying close to a person may cause fear and gives surely no chance to detect details from the body of these bats. Their hidden way of life poses also big problems for researchers to investigate their life cycle, morphology, their potential vectorship for agents of diseases, etc. Thus, the recent knowledge on bats is rather poor compared to that on other zoological genera. Since former generations of humans had much less possibilities to get information on these nocturnal animals and the strong belief that during the night only malicious diabolic individuals of any kind fly around, myths on bats were created and transferred orally from generation to generation reaching until today in some countries.

Thus, this book throws some glimpses in ten different chapters on the different features of bats covering their morphology, their biology, their potential vectorship of agents of diseases, and their own parasites and on similarily looking parasites. Furthermore, special attention is given at the numerous myths that had been told or are still told on these unique organisms. This all together is presented to all who are interested in natural and especially in biological phenomena.

Frankfurt, Germany — Sven Klimpel
Düsseldorf, Germany — Heinz Mehlhorn

Acknowledgement

The quick, careful, and attractive publication of so many data based on own research and/or selected from papers of a very broad spectrum of international authors is not possible without the help of many persons. At first we thank all contributors for their in-time delivery of the manuscripts, so that latest research aspects could have been included. Then we are indebted to the efforts of all co-workers of each group who gave text and figures their final shape. Our thanks are also directed to Mrs. Anette Lindqvist from Springer and the production team of SPi Global. Their duly and competent efforts made it possible to present these comprehensive insights into the present knowledge of a unique group of flying mammals, the importance of which is surely still underestimated despite its world-wide presence and its endangered survival.

Frankfurt, Germany — Sven Klimpel
Düsseldorf, Germany — Heinz Mehlhorn
June 2013

About the Editors

Sven Klimpel studied Biology in Kiel and then completed his doctorate and his habilitation at the Institute of Zoomorphology, Cell Biology and Parasitology in Düsseldorf. Since 2010 he is professor and head of the department "Medical Biodiversity and Parasitology" at the Biodiversity and Climate Research Centre (BiK-F) of the Goethe-University (GU, Institute of Ecology, Evolution and Diversity) and the Senckenberg Gesellschaft für Naturforschung (SGN). His primary research interests are the ecology, evolution, and host–parasite coevolution of aquatic (limnic, marine) and terrestrial protozoan/metazoan parasites and pathogens (e.g., viruses, bacteria, fungi). He is particularly interested in how changes in environmental conditions influence dispersal and migration of parasites/pathogens and their intermediate hosts/vectors. His work integrates a variety of disciplines, such as phylogenetics, systematics, ecology, ichthyology, zoogeography, oceanography, genomics, and population genetics. In his lab, he and his coworkers combine traditional morphological methods with up-to-date molecular techniques. He participates in numerous scientific research cruises, including some to the Southern Ocean (Antarctica).

Heinz Mehlhorn Düsseldorf, Germany. He has investigated the transmission pathways of human and animal parasites for over 40 years at German and international universities and he and his university spin-off company Alpha-Biocare have developed many antiparasitic medical products based on more than 20 patents—several in cooperation with big international companies. He has published 25 books, more than 250 original papers, and has served as Managing Editor of the journal Parasitology Research since 1981. Many renowned international scientists did their PhD work in his laboratory and remain still today interconnected as a large group of lovers of parasitology.

Contents

Chapter 1
Introduction: The World of Bats

Heinz Mehlhorn

Abstract Bats are unique flying mammalians that occur worldwide feeding either on insects, fruits or even on blood depending on the species. Since they are active during the night, many myths have been launched during the last 500 years. All these facets are shortly considered here and prepare for a hopefully intense look into their hidden life.

Key words Bats • Chiroptera • Vampire bats • Vectors of diseases • Chupacabras • Batman • Mammalia • Ultrasound • Parasites • Flies • Lice • Louse flies • Sheep milker • Blood sucking • Morphology • Reproduction

Bats—their English name points to their fluttering up and down wing movements—are a peculiar group of the so-called Mammalia (= animals with breast nipples) in the Zoological System, where they are described as Chiroptera (*Greek*: cheir = hand, pteron = wing) (Neuweiler 1993; Niethammer and Krapp 2004, Claus et al. 1932). This term refers to their morphological peculiarity that thin portions of leathery skin (Plagiopatagium, Propatagium, Dactylopatagium and Uropatagium) are stretched between the body and all fingers reaching from there to the knuckles of both feet (Claus et al. 1932; Fig. 1.1). These skin portions can be folded, when the bats rest in their typical position: head down under (Figs. 1.2 and 1.3), fastholding by help of the claws and their five toes at each foot. Their size may vary considerably between the different species. There exist rather large species (e.g. *Macroderma gigas* in Australia), which reach a body length of up to 20 cm, show a "wingspan" of up to 60 cm when stretched and weigh nearly 200–250 g. The smallest species (e.g. *Craseonycteris thonglongyai*) have a body length of about 3 cm at the maximum and a body weight of only 2 g thus belonging to the smallest

H. Mehlhorn (✉)
Institute for Parasitology, Heinrich Heine University, Universitätsstr. 1, 40225 Düsseldorf, Germany
e-mail: mehlhorn@uni-duesseldorf.de

S. Klimpel and H. Mehlhorn (eds.), *Bats (Chiroptera) as Vectors of Diseases and Parasites*, Parasitology Research Monographs 5, DOI 10.1007/978-3-642-39333-4_1,

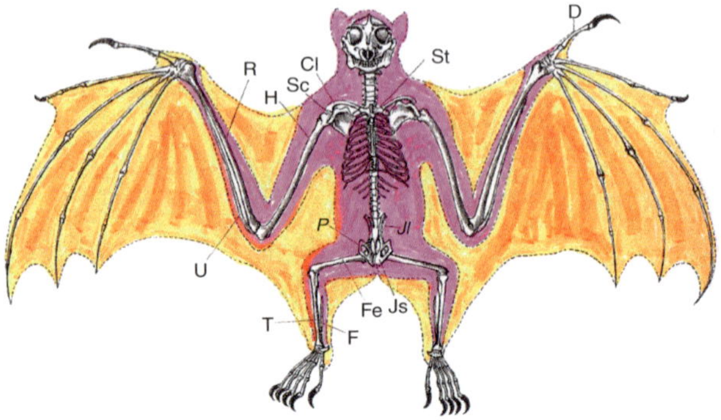

Fig. 1.1 Diagrammatic representation of the body "wings" and skeleton of the chiropteran species *Pteropus* sp. according to Claus et al. 1932. Note that at the thumb and the index finger of each hand claws are present as well as at each of the five toes. Cl = clavicula; D = thumb, pollex; F = fibula; Fe = femur; H = humerus; Il = Os ileum; Is = Os ischii; P = Os pubis; R = radius; Sc = scapula; St = sternum; T = tibia; U = ulna

Fig. 1.2 Australian "flying fox" hanging head down at a branch of a tree in the Botanical Garden of Sydney—happily sleeping during daytime

mammalians at all. The body of all bats is covered with very fine, shiny (often brownish) hair. Bats are active at night seeking their food, which consists—depending on the species—of insects, fruits or even of blood. Since their rather quick flight is absolutely silent, so that they may approach and pass humans like a dark flash, many fairy tales developed in different human cultures (Chaps. 7, 8). Thus, their hidden way of life caused fear, especially in those cases, when many bats fly simultaneously around houses or around heads of persons, since they live together in groups of often more than 100–150 animals hiding themselves during daytime in the treetops (e.g. plant eating flying foxes e.g. in Australia) or in tunnels of railways, galleries of closed mines, in old, rottening empty buildings on practically all continents except for Antarctica. This peculiar way of life and their strange behaviour at night led to the fact, that for a long time the knowledge on their

Fig. 1.3 A "flying fox" being disturbed during sleeping showing his pointed teeth to announce its "annoyance"

morphology and their mode of reproduction was scarce. Thus it was a big surprise when it was discovered, that these animals give birth to their offspring and do not lay eggs at hidden places. In general, they bear only one "baby" per year, although they possess two nipples at their breast. During the flight of the mother, the young bats are transported being attached to the hair of mother's belly often biting into one of the nipples for further stabilisation during the flights.

Absolutely surprising was also the discovery that the bats support their nightly flights not only by excellent eyes, which can catch even remnants of light, but also that they have developed a system of expelling ultrasound waves (Holland et al. 2006; Skiba 2009), a system of perception of the repelled waves and in addition they possess a nerve system, which allows them to react and to steer their muscles in milliseconds in order to avoid crashes with invisible obstacles in the night. These inventions are not unique, however, extremely scarce in the animal kingdom. It is less than 100 years that humans detected these waves, which reach far beyond the border of human hearing ranging up to 200 kHz. Thus the bats are extremely interesting with respect to their abilities and activities. While bats in warm countries are active all year around, the European species hibernate.

On one side most of the bats—especially the insect feeders—are very useful for humans and thus are strictly protected, since their breeding places are endangered,

Fig. 1.4 Representation of the title page of the famous book of the English writer Bram Stoker that appeared 1894 and made vampirism famous (in the sense of bloodsucking immortal humans that stay during daytime in their coffins and suck blood at night with the effect that their victims become vampires)

especially in the surroundings of the growing industrial cities. On the other side, some of the species damage fruit plantations by eating fruits or just by biting into fruits in monocultures in Australia and Southeast Asia (Figs. 1.2 and 1.3). A few bat species in South and Central America (Chap. 7) even suck blood of animals staying on meadows at night. These hosts are injured by the sharp and pointed teeth of the vampire bats, suffer from blood loss and even may become infected by pathogens such as those of true rabies (Lyssa virus). Indeed there exist several more of such bat transmitted viruses. Therefore, also the harmless and very useful species that feed on insects have recently been set in the focus of scientific interest, since they are now known also as possible true or mechanical vectors of severe agents of diseases (especially of viruses) that may hit humans and animals (see Chap. 3), if they get in contact with bats or their infected faeces (Drosten 2012). Especially Drosten's laboratory and the microbiological institute of the University of Hong Kong showed that numerous bat species were carriers of the human-associated Corona virus (SARS-CoV). In total, eight variations had been characterised (http://covdb.microbiology.hku.hk:8080/COV-newpages/index.html).

Thus, this book is designed to present some interesting and important features on the biology, morphology, reproduction and the potential vectorship of bats on one side and to throw a glimpse on the mysterious beliefs that are still whispered today when bats are flying (Fig. 1.4).

Thus this book comprises several chapters of different length that deal with some of the most interesting and most important findings in bat research. However, it is evident that there remain many more surprising news to be detected in the future.

References

Claus C, Grobben K, Kühn A (1932) Lehrbuch der Zoologie, 10th edn. Springer, Berlin

Drosten C (2012) Unbelievable reservoir of viruses. http://www.uni-bonn.de

Holland RA, Kirschvik JL, Doak TG, Wikelski M (2006) Bat navigation: bat orientation using Earth's magnetic field. Nature 1112:116

Neuweiler G (1993) Biology of Chiroptera. Thieme, Stuttgart

Niethammer J, Krapp F (eds) (2004) Handbook of the mammalia in Europe. Chiroptera, vol 4/2. Aula, Wiesbaden

Skiba R (2009) Fledermäuse: Kennzeichen, Echoortung und Detektoranwendung, 2nd edn, vol 648. Brehm-Bücherei

Chapter 2
Bats: A Glimpse on Their Astonishing Morphology and Lifestyle

Volker Walldorf and Heinz Mehlhorn

Abstract Bats (*zool.* Chiroptera) are unique in the zoological group of Mammalia with respect to their morphology, their lifestyle, their capacity to fly and to orientate themselves during the darkest nights and by the fact that they succeeded in extending their biotopes over all continents except for Antarctica and the Northern arctic regions. Due to their nocturnal activity and their resting, respectively, sleeping during daytime at hidden places, information on this group remained scarce compared to other mammalians. Thus this chapter aims to throw some short glimpses on their morphology and on some other astonishing peculiarities.

Keywords Morphology • Megachiroptera • Microchiroptera • Pteropodidae • Echolocating • *Pteropus* • *Acerodon* • *Craseonycteris* • *Desmodus* • Monophyly • Mammalia • Wings • Flying foxes • Vectorship of viruses • *Pterosauria* • Patagium

2.1 Systematic Position

The order Chiroptera is subdivided into the Megachiroptera (Old world fruit bats, Flying foxes, "Megabats") and the Microchiroptera (Echolocating Bats, "Microbats") and is one of the most successful and abundant mammalian groups. The Megachiroptera are represented by only one family, the Pteropodidae, whereas the Microchiroptera comprise 16 families. Within the Pteropodidae, 42 genera and 166 species are described; 16 families of the Microchiroptera include 135 genera with 759 species (Simmons 2005; http://www.ucmp.berkeley.edu/mammal/eutheria/chirolh.html).

V. Walldorf
Institute for Zoomorphology, Cytology and Parasitology, Heinrich Heine University, Universitätsstr. 1, 40225 Düsseldorf, Germany

H. Mehlhorn (✉)
Institute for Parasitology, Heinrich Heine University, Universitätsstr. 1, 40225 Düsseldorf, Germany
e-mail: mehlhorn@uni-duesseldorf.de

S. Klimpel and H. Mehlhorn (eds.), *Bats (Chiroptera) as Vectors of Diseases and Parasites*, Parasitology Research Monographs 5, DOI 10.1007/978-3-642-39333-4_2,

System of Chiroptera	
Class	Mammalia
Subclass	Higher Mammalia (Eutheria)
Order	Chiroptera
1. Suborder	Microchiroptera
Family	Rhinopomatidae
	Natalidae
	Noctilionidae
	Rhinolophidae
	Molossoinidae
	+10 further families
2. Suborder	Megachiroptera (Old world fruit bats, flying foxes)
Family	Pteropodidae

Most of the Microchiroptera do not reach a weight of 30 g and are significantly smaller than the Megachiroptera. Nevertheless, there are some microbats exceeding the dimensions of the smallest Megabats. For example in Australia, the so-called ghost bat (*Macroderma gigas*) reaches a wing span of up to 60 cm and a weight of 200 g at the maximum. The largest bats can weigh up to 1,500 g and may reach a wingspan up to 1.7 m (Megachiroptera, genera *Pteropus* and *Acerodon*), whereas the smallest representative has a wingspan of only 15–17 cm and reaches a weight of 1.5–3 g (Microchiroptera, *Craseonycteris thonglongyai*) (Westheide and Rieger 2010).

The ancestors of the Pteropodidae had apparently been capable of echolocating, too. But this ability was lost in most members of this family during evolution. Simultaneously their efficiency in night vision was improved, as it becomes obvious in formation of bigger eyes compared to those of the majority of the Microchiroptera (Westheide and Rieger 2010).

The monophyly of the group has been doubted and it was supposed that Megachiroptera and Microchiroptera might have developed independently from different ancestors (Jones and Genoways 1970; Pettigrew et al. 1989). This would as well imply the convergent evolution of the active flight as other similarities within the two groups. However, the results of numerous studies now strongly support the monophyly of the members of the order Chiroptera (for detailed bibliography, see Simmons and Conway 1997; Simmons et al. 2008).

However, the systematical belongings of the Chiroptera still are under discussion. There had been established different phylogenetical trees depicting the relationships between the groups of the Chiroptera. Within one tree, the group is subdivided into only five taxa. The Pteropodidae are classified as a sistergroup of the Rhinolophidae, Emballonuridae, Noctilionidae, and Vespertilionidae, which represent the former Microchiroptera (Smith 1976; Westheide and Rieger 2010).

The majority of the established determination keys utilize the subdivision of the Chiroptera into Mega- and Microchiroptera. This classification is therefore used in the present text. The Chiroptera are the only vertebrates besides the formerly existing pterosauria and the recent birds that have evolved the ability of an active

Fig. 2.1 *Pteropus seychellensis* during flight

flight. Furthermore, they are the only actively flying mammalians (Fig. 2.1). Other Mammalia (as for instance the gliders) are only able to fly passively by gliding (Simmons and Conway 1997). In contrast to the gliders, all bats sleep or rest in a head down-under position (Figs. 2.2 and 2.3). The oldest known representative of the group of bats derives from the Eocene (about 50 million years ago) (Lawlor 1979; Carroll 1988).

2.2 Morphology

All bats show the characteristics of mammalians, as for instance, hairs, lactiferous glands, three ossicles, and the ability to maintain their body temperature (Maywald and Pott 1988).

In contrast to the wing of a bird, which is characterized by strong arm bones, two fingers, and feathers, the wings of bats consists of rather thin arm bones and four strongly elongated bones of the hand that strengthen the wing membrane (patagium, Figs. 2.4, 2.5, and 2.6). This organization led to the name of the group: Chiroptera = hand wing (from *greek*: cheir = hand, pteron = wing). The short clawed thumbs are not integrated into the wing membrane but are freely movable and used to grab, to hold, and to climb. The bones of the hand, which comprise four metacarpalia and elongated phalanges, strengthen the wing membrane (Figs. 2.1, 2.4, and 2.5), which stretches between the bones of the arms, the hands, the side of the body, and the bones of the hind limbs. The latter are rotated by nearly 180°. That causes the knees to be directed upwards. There is a chondral spur at both hind limbs, which in some species may be bony (osseous). It is called calcar and serves as support of the rear end of the wing membrane, the uropatagium. All toes are free and provided with laterally flattened claws and a locking mechanism. It consists of tendons and their

Fig. 2.2 *P. seychellensis* just landed at its resting place

Fig. 2.3 *P. seychellensis* in typical sleeping position. The five toes at the feet are clearly visible

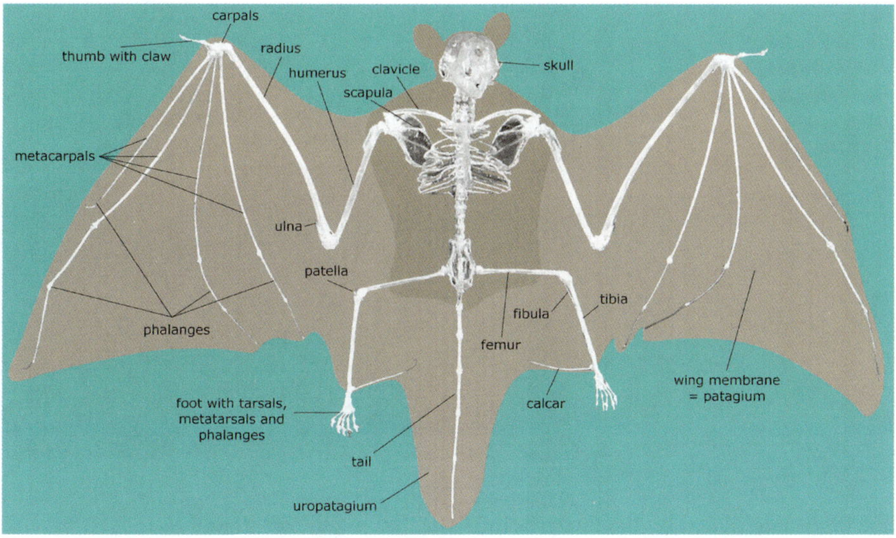

Fig. 2.4 Semiphotographic and diagrammatic representation of the skeleton of a microchiropteran bat to show that the wing bones originate mainly from the hand. The wing membrane and body shape are schematically represented

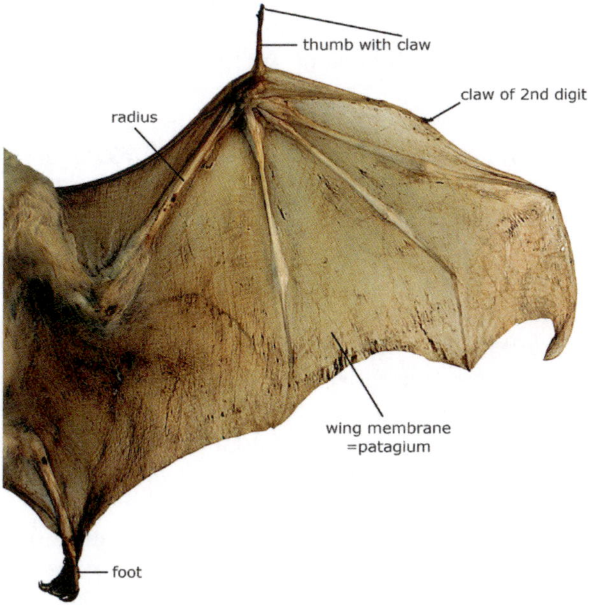

Fig. 2.5 Photograph of the ventral side of the left wing of a megachiropteran bat. This Pteropodidae bat species exhibits besides the clawed thumb, that all microbats have, another claw at the second digit

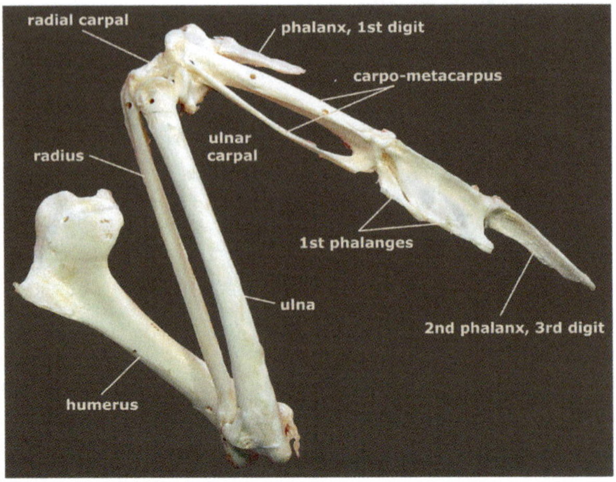

Fig. 2.6 Photograph of a bird wing skeleton with extremely reduced or eradicated hand bones

sheath being locked by the bat's weight, thus allowing the animal to rest at its roost without using a muscle (Maywald and Pott 1988).

The wing membranes consist of connective tissue, elastic filaments, muscles, nerves, and blood vessels (Fig. 2.1). It is covered on either side by a thin epidermal layer. The elastic fibers contract the relaxed wing, thus reducing its surface. During the flight, the muscles keep the wing membrane tightened. Contractile vessels assure the supply of the wing margin with blood. During flight, the wing membrane with its blood vessels serves as cooling surface to control the body temperature (Maywald and Pott 1988).

The flight muscles of birds are located in their chest, whereas the upstroke muscles of bats are located on their back and the downstroking muscles are attached to a keel at the breastbone, similarly as in birds (Fenton 1992). The shape of chiropteran skulls reveals an adaption to their diets (Fig. 2.7). For instance, the skulls of nectar feeding species are slim and elongated, allowing easy access to the nectar in the blossoms. Frugivorous and insectivorous species often have relatively short and blunt skulls (Hill and Smith 1984; Westheide and Rieger 2010) (Fig. 2.7)

The dentition shows adaptation to the diet, too (Fig. 2.7). The dental formula originally contained 38 teeth. However, the dentition of the recent species is extremely variable and the number of teeth has been reduced even to 20 in the vampire bat (*Desmodus rotundus*). There exist altogether about 50 different dental formulas within the Chiroptera (Westheide and Rieger 2010). About 75 % of the Chiroptera are insectivorous feeding mainly on beetles and moths. About 20 % of the species are frugivorous (Pteropodidae and Phyllostomidae) and only a few species feed on nectar and pollen (Pteropodidae and Phyllostomidae). Less than 2 % are predators of small mammals, birds, reptiles, amphibians, and fishes. In addition, there are only three species that feed on blood of vertebrates (see Chap. 7).

Some specializations of the alimentary tract have been developed as adaptation to the kind of food. Nectar feeding bats possess long tongues that can be protruded to ingest food that is deeply hidden in plant blossoms. The esophagus of

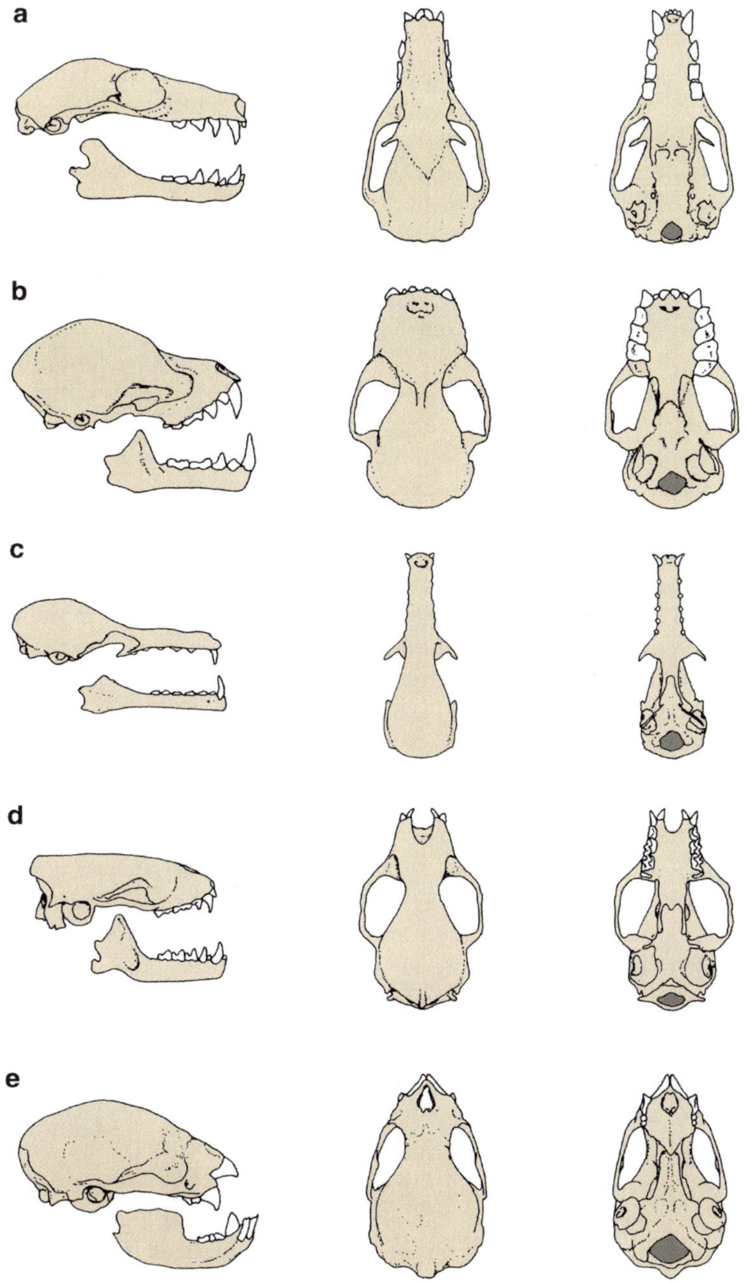

Fig. 2.7 Variety of chiropteran skulls represented in lateral, dorsal, and ventral sight to show the adaptation of skull shape and teeth formula to the diet of the species. (**a**) Megachiroptera, Pteropodidae (flying fox), *Epimorphosis* sp., fruit-feeder. (**b**) Microchiroptera, Phyllostomidae, *Artibeus* sp., fruit-feeder. (**c**) Microchiroptera, Phyllostomidae, *Choeronycteris* sp., nectar-feeder. (**d**) Microchiroptera, Vespertilionidae, *Eptesicus* sp., insectivorous. (**e**) Microchiroptera, Phyllostomidae, *Desmodus rotundus*, hematophagous. The proportions between the skulls do not correspond to the reality (Modified after Lawlor 1979)

insectivorous bats is lined by a horny epithelium. Bloodfeeding species, such as *Desmodus rotundus*, are provided with a long, dilatable blind sac to store great amounts of blood. In general, the intestine of insectivorous bats is shorter compared to the longer ones of frugivorous species (Westheide and Rieger 2010).

The sensory organs of bats are also well adapted to the "way of life" of the different chiropteran groups. As far as known, the olfactory sense is comparatively well developed in fruit-, nectar-, and bloodfeeding bats but is less good in insectivorous species. The eyes of the members of the Megachiroptera are essential for orientation and foraging of most species and therefore are in general very large and thus efficient even in dim light. Orientation of flying foxes in the night is accomplished by optical and olfactory senses. In contrast, the eyes of the echolocating Chiroptera (Microchiroptera) are mostly small but help to catch the prey, too. Other microchiropteran bats, such as Megadermatidae (False vampire bats), Phyllostomidae (Leaf-nosed bats, e.g., the hematophagous *Desmodus rotundus*), and several members of the Vespertilionidae (Vesper bats; Vespertilio= bat, *latin* from vesper = evening), have developed larger eyes (Hill and Smith 1984; Westheide and Rieger 2010).

The ears of flying foxes (Old world fruit bats) do not possess special features. However, ears of echolocating bats are often characterized by striking and variable formations of the morphology of the outer ear as well as of the middle and inner ear. Large auricles are found in Microchiroptera producing sounds at lower frequencies, while the smaller auricles of some species send out higher frequencies. Sounds—especially in the ultrasonic range—are produced within the larynx and are emitted through the openings of mouth and nose. The echos are recorded by help of the ears. The used frequencies range from 8 to 220 kHz and the sending lasts 2 ms to more than 100. The phonetic pattern can be modified and adjusted to the prevalent conditions. The nose and lips of bats can be modified extremely for instance by noseleafs, folds, and wrinkles. Function and purpose of most of those modifications are not understood. The nose of the vampire bat (*Desmodus rotundus*) is surrounded by so-called pit organs, which are sensitive for infrared radiation and allows the vampire bat to detect its bloodsource by the emitted body heat (Gracheva et al. 2011; Westheide and Rieger 2010).

2.3 Food of Bats

The Old world fruit bats (Megachiroptera and Pteropodidae) mainly feed on fruits. Some live on blossoms, pollen, or nectar. Especially the fruit feeding species definitely can introduce severe damages in fruit plantations. On the other hand, several plants need to be pollinated by other fruit bats (Grzimek 1975). At least 289 plant species worldwide (e.g., the African Baobab tree and the Australian Iron tree) depend on propagation by flying foxes. This can only be brought about by large populations of these animals (Fujita and Tuttle 1991). Besides fertilization of plants they play an important role in spreading and transporting the seeds of plants into new areas. Many of the Microchiroptera act in the same manner as flying foxes as transporters and pollinators for many plant species, for instance, guavas, avocados, mangos, agaves, bananas, etc. (Grzimek 1975; Hill and Smith 1984; Fujita and Tuttle 1991; Nowak 1991).

Microchiroptera have made accessible very different food sources. Most of them feed on insects. In turn, others feed on fruits, blossoms, pollen, or nectar as do the Megachiroptera, others catch little frogs or even fish (e.g., bats of the genus *Myotis*).

Microbats play a fundamental role within the ecosystems, since the insectivorous species control certain insect populations. The abundance of certain insects would be much higher without this regulation. Some bat species for instance feed especially on mosquitoes, which are known to be able to transmit the agents of severe diseases in animals and humans. One single bat as the Little brown bat can catch up to 600 mosquitoes per hour (Grzimek 1975). Extrapolation of a quantity of 500 mosquitoes per bat per day by 7 days and a total population of 1,000 bats adds up to five million mosquitoes per night (Callisher et al. 2006). The latter example demonstrates the fundamental role of bats in pest control. Some few species in South and Central America feed on blood. After scratching the skin by help of their pointed and sharp teeth they lick the flowing blood (see Chap. 7).

2.4 Social Behavior and Reproduction

Bats are socially highly developed mammals. Many of them live in groups. Some species may create giant groups with up to one million of individuals. Species like the so-called Mexican free-tailed bat can aggregate up to 300 individuals per square meter (Constantine 1967). In contrast, there are other species, which live solitary most of the time (Wund and Myers 2005).

On daytime the bats rest in trees, caves, buildings, etc. (Figs. 2.2 and 2.3). Some bats of the temperate regions can fall in torpor daily and hibernate in cold seasons (Lyman 1970). Other species may migrate yearly over long distances (up to 800 miles) to the sites, where they overwinter (Cockrum 1969).

A highly developed social behavior has been observed in the common vampire bat (*Desmodus rotundus*) in South America. As this bat dies of starvation after only two nights without a blood meal, regurgitation and sharing of blood with other starving individuals regularly takes place (Wilkinson 1990).

Most bats are polygyn, so that males establish a harem. For several bat species, courtship behavior has been observed and female bats may aggregate on nursery roosts. The bats of the temperate zones breed only once per year, but in warmer climates some species can give birth to several babies, but at each time only one. Chiroptera may reach an extraordinarily long life span of often 20 years. For a *Myotis* species, even a life span of 40 years had been reported (Westheide and Rieger 2010).

2.5 Geographic Range

Bats belong to the most widely distributed land mammals, only humans and rodents have reached a higher prevalence rates (Wimsatt 1970). The 170 species of Megachiroptera (being included in the single family Pteropodidae) are found exclusively in the tropical zones of the Old world. That is Africa, Indochina, East Indies, Australia including the islands north and east of the continent. However, they do not occur in the Americas. On the other hand, the members of the second suborder Microchiroptera have spread all over the whole world with the exception of mountain peaks, Arctic and Antarctic regions, and some isolated islands. Therefore, it is understandable that the different groups of the Chiroptera have attained the highest diversity of species in the tropical regions.

2.6 Parasites

Bats are hosts for a large number of endo- and ectoparasites. Endoparasites are protozoans such as instance *Plasmodium* and *Trypanosoma* species. For platyhelminthes (trematodes and cestodes) as well as for nematodes, bats may serve either as intermediate or as final host. Ectoparasites that live on bats belong to the groups of ticks and mites. Other parasites are insects. Especially the groups of bugs (Hemiptera), fleas (Siphonaptera), and flies (Diptera) are very common. One family of flies (Streblidae) is extremely specialized as some of their members have lost their ability to fly during evolution and live now as "bat lice" in the fur of bats (Wund and Myers 2005).

2.7 Bats as Victims and Vectors of Diseases

More than 70 viruses have been isolated from bats (Callisher et al. 2006). They can be transmitted within the large populations of bats between the individuals of one species. Since often several different bat species roost in the same habitat, intraspecific transmission takes place too. In addition, these viruses can be exchanged easily between bat species living relatively constant in their habitat and other species, which migrate as groups. Viruses as for instance the rabies virus group can cause serious illness in some bats and often lead to death (Boxes 2.1, 2.2, and 2.3). However, other bats may overcome the infection and seem to store the virus in their body. Field studies revealed that bats when infected with rabies may become aggressive against other bats or against other animals in their surroundings. Thus, bites become a possible way of interspecific transmission of the virus (Bell 1980). Several of these "bat"-viruses are known to be transmitted to other mammals including humans as it is shown in the following examples.

Several representatives of the family of Rhabdoviridae (genus *Lyssavirus*) have been found in tissues and salivary glands of bats and had been grouped as "bat *Lyssa* viruses." Furthermore exist strains of *Lyssa* viruses, which are known to induce the typical Rabies disease in humans and many animals (see Box 2.1). Bat *Lyssa* disease and typical rabies can be transmitted within the populations as well as from bats to other mammals by biting, scratching, and by aerosols as has been documented in bat populations roosting in caves (Callisher et al. 2006; Constantine 1967; Winkler 1968).

Talking of rabies often associates the vampire bats and transmission of rabies during their blood uptake. The saliva of the vampire *Desmodus rotundus* has been demonstrated to contain rabies viruses, after the bat has overcome the illness. There are reports on several death cases of humans living or temporarily staying in Latin America—where this bat is to be found. They obviously had been infected by the bites of *Desmodus rotundus* (MMWR 2011; Callisher et al. 2006). Of the three known vampire bat species only *Desmodus rotundus* has been witnessed to feed on mammalian blood (Belotto et al. 2005) besides numerous other hosts inclusive non mammalians. Rabies within populations of the vampire bat *Desmodus rotundus* is a potential danger to livestock (particularly for cattle) in Latin America (Batfacts, http://www.si.edu, 1980). Rabies virus is transmitted from bat to bat by saliva during bites (Tuttle 1990; Brass 1994). In several countries of Latin America, vaccination campaigns are carried out to protect the cattle. The vaccination together with efforts to lessen the vampire populations led to a significant reduction of rabies infection in some countries (Arellano-Sota 1988). Not only vampire bats but also other bat species can transmit rabies to various other mammals, which then may infect other mammals including humans (Callisher et al. 2006).

As is outlined in a review of human rabies cases in Latin America transmitted by vampire bats, the outbreaks of this disease are obviously strongly influenced by several factors—biological and nonbiological ones (Schneider et al. 2009).

Biological factors amongst others are the presence of:

1. Vampire bats
2. Shelter and roosting sites for the bats
3. Food sources
4. Reservoir of rabies virus

Nonbiological factors for example are:

1. Type of human activities in the area
2. Changes in activities of animals in the area
3. Changes of the environment caused by activities such as productive processes, working and live style, etc.

These factors altogether for instance could lead to the situation that vampire bats, which lost their food sources by human activities, start to attack people and try to use them as blood source.

Another representative of this virus group is the Australian-bat-*Lyssa*-Virus (ABLV). It can persist in populations of flying foxes namely in several *Pteropus* species and several Australian species of the Microchiroptera (Callisher et al. 2006). Up to now, this virus has caused three human death cases in Australia. In 1996, the *Lyssa*-virus was detected in a black flying fox (*Pteropus alecto*). Within the same year, a woman in Queensland became ill and died. Investigations of samples confirmed that she was infected by *Lyssa*-virus. The same occurred in 1998, when a woman in Northern Queensland became ill and died. She was infected with *Lyssa*-virus, too. Research showed that some bats had *Lyssa*-viruses inside their salivary or in their saliva glands. This virus can be transmitted to humans by bites or scratches of infected fruit bats. (http://www.csiro.au/en/Organisation-Structure/Divisions/Animal-Food-and-Health-Sciences/Infectious-diseases-overview/Australian-bat-lyssavirus.aspx; Constantine 1962).

In 2013, the third fatal case of death by infection with the ABLV virus-type (Box 2.1) occurred in Australia. A 8-year-old boy died, who obviously had been bitten by a bat 3 months before during holiday on the White Sunday Islands—a surfer's paradise along the Barrier Reef (Francis et al. 2013).

The Hendra virus belongs to Paramyxoviridae, genus *Henipavirus*. It was detected in Australia in 1994 for the first time. Horses and humans became infected and died. However, humans were not infected directly by infected bats but had close contact to infected horses. In the years 1994–2010, several humans were infected and died after infection with the Hendra virus, while 14 clusters of virus infections were recorded in horses. In 2011, in the East Australian states of Queensland and New South Wales, 18 outbreaks and 24 cases in horses were recorded. In this year, the first infection of a dog was documented. This virus is apparently able to infect different animal species. Fruit bats are believed to be the natural "host" of this virus. They carry the virus, but the virus does not affect them severely (Callisher et al. 2006; http://www.csiro.au/Outcomes/Food-and-Agriculture/Hendra-Virus/Research-findings.aspx). Both a pre- and a postexposure vaccination are available against the typical rabies virus and ABLV-type as well. There is still no vaccination against the hendra virus, although research is underway since long (Hendra Virus Infection 2012; Rabies 2012; Australian Bat Lyssavirus 2013).

Box 2.1: Classification of *Lyssa*-Viruses (According to Neumeister et al. 2009)

Genotypes	Virus	Potential/real hosts
1	Rabies (RABV)	Carnivores, bats
2	Lagos-Bat (LBV)	Fruit feeding bats
3	Mokola (MOKV)	Unknown
4	Duvenhage (DUVV)	Insect feeding bats
5	European bat-*Lyssa* (EBLV 1)	Insect feeding bats
6	European bat-*Lyssa* (EBLV 2)	Insect feeding bats
7	Australian bat-*Lyssa* (ABLV)	Fruit and insect feeding bats
Not yet classified	Aravan virus (ARAV)	Insect feeding bats
	Khujand virus (KHUV)	
	Irkut virus (IRKV)	
	West Caucasian bat virus (WLBV)	

Box 2.2: Description of the Viruses

Rabies viruses belong to the family of **Rhabdoviridae**

Size	100–300 nm in length and 75 in diameter
Shape	Bullet-like
Cover	G-proteins with spikes
Contents	Helically arranged nucleocapsid containing single stranded RNA of about 12,000 nucleotides with a negative polarity coding for 5 viral proteins

Box 2.3: Symptoms of Human *Lyssa*-Disease

Incubation period	Variable, mostly below 30 days up to 90 days or even years
Prodromal stage	Headache, vomiting, fever for 2–7 days
Neurological phase	So-called "wild wrath," aggressivity, confusion, delusions, aero-, hydrophobia and hypersalivation, paralytic phases, death due to circulatory collapse
Diagnostic measurements	Immunofluorescence test, immunochemical methods, virus isolation; primers are available for amplification of all genotypes

2.8 Bats and Men

Many myths and misunderstandings are linked to the bats (Fig. 2.8). Their silent flight frightened people and was thought to be a portent in the ancient world (Harenberg 1733). Especially the ability to fly in the darkness was a mystery. It was inapprehensible that a creature half mouse, half bird could fly in the night, when the eyes of humans and other diurnal living beings lost totally their efficiency (Figs. 2.9 and 2.10). Thus it was believed that bats must possess magical powers or to be in league with the devil. Possibly this was one reason to illustrate the devil and his cronies with the wings of bats, whereas the angels got white bird wings (Maywald and Pott 1988).

The imagination of so-called vampires—dead humans who suck blood at night—existed long before the "real" vampire bat (*Desmodus rotundus*) had been discovered and was first described as *Phyllostomus rotundus* (Geoffroy 1810) (see Chap. 8).

For long the important role bats that they play within the ecosystems was completely misunderstood or even completely unknown. The threat that they could transmit rabies disease as well that they accomplish damages in fruit plantations led to attempts to eradicate or to reduce bat populations. Especially eradicating campaigns against vampire bats killed numerous other bats species, which only looked similar like vampires or lived in the same caves. These efforts, however, did not really reduce the transmission of rabies (Callisher et al. 2006) but

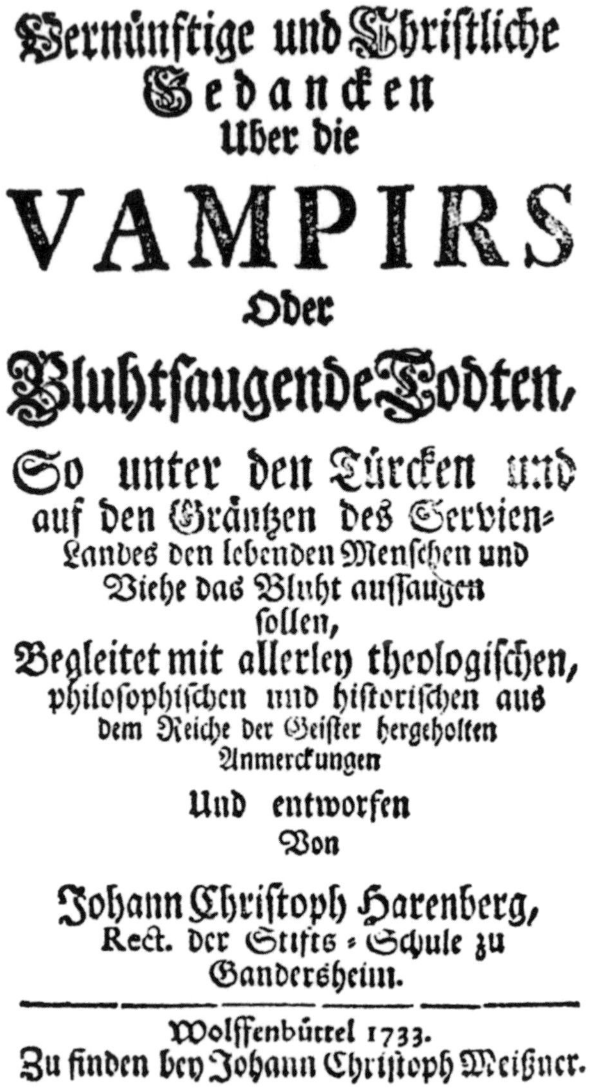
Vernünfftige und Christliche
Gedancken
Uber die
VAMPIRS
Oder
Bluhtsaugende Todten,
So unter den Türcken und
auf den Gräntzen des Servien-
Landes den lebenden Menschen und
Viehe das Bluht aussaugen
sollen,
Begleitet mit allerley theologischen,
philosophischen und historischen aus
dem Reiche der Geister hergeholten
Anmerckungen
Und entworfen
Von
Johann Christoph Harenberg,
Rect. der Stifts-Schule zu
Gandersheim.
Wolffenbüttel 1733.
Zu finden bey Johann Christoph Meißner.

Fig. 2.8 Title page of Harenberg's book on vampires (1733). The German text says: Reasonable and Christian thoughts on the vampires or bloodsucking dead humans, which are said to suck off the blood of living humans and cattle resident in countries at the Turkish border and at the borders of Serbia, accompanied by several theological, philosophical and historical comments, which have been taken out of the kingdom of ghosts and composed by Johann Christoph Harenberg, rector of the monastery-school at Gandersheim. Wolfenbüttel 1733. To be found at Johann Christoph Meißner

introduced bad effects on biotopes and on plantations. In addition, other bat populations suffered severely from these measurements and their number was diminished by the extensive use of insecticides amongst others DDT. In particular DDT affected the bats and their offspring in Europe and the USA in the 1960s (http://www.earthlife.net/mammals/chiroptera.html).

From the human point of view, bats may exhibit a potential danger as a reservoir of numerous viruses. But the number of human death cases due to bat-derived fatal diseases is very small compared to those caused by other infectious diseases (mosquito- or tick transmitted ones). As for infections of cattle with rabies virus,

Fig. 2.9 Gray-headed flying foxes, *Pteropus poliocephalus*, resting in trees in the Botanical Garden Sydney, Australia

Fig. 2.10 Gray-headed flying foxes, *Pteropus poliocephalus*, resting in trees in the Botanical Garden Sydney, Australia – note the large numbers crow-wing together

it has to be considered, that obviously high numbers of infection and death cases eventually have been created—at least in part—by the dramatic changes of the environment. Forest clearings in order to get grasslands drew off the habitats of bat populations together with their food resources and therefore the bats entered the environment of cattle and humans. These large numbers of cattle offered large amounts of food for bats and thus increased the growth and denseness of vampire populations. The latter abets a higher infection level of bats directly with rabies and thus brought about the increase of rabies cases in cattle—a constantly running "circulus vitiosus."

Destruction of bat habitats followed by reclaiming land and settlement of people could raise the number and closeliness of contacts between bats and humans, too. Especially in cases when bats used buildings as new shelter and resting places. Trials to eradicate bats—especially vampire bats—by destruction of their roosting places and by deployment of agents such as coumarin being applied on cattle limited the numbers of all bats.

Vaccination of cattle as well as downsizing the droves appears to be a better attempt in controlling rabies. In Europe, vaccination of red foxes against rabies was successfully started in Switzerland 1978. In 1983, Germany succeeded with its vaccination program and stopped rabies practically everywhere in recent times. In the following years, many other European countries did the same and in 2008 rabies was eradicated or controlled in most countries of Western and Central Europe (http://www.who-rabies-bulletin.org/about_rabies/Control.aspx). Lyssa viruses within the reservoir host bats, however, were not affected by this campaign.

Although it is utopian with respect to the increasing world population, that it will be possible to reduce the livestock breeding sites and to renaturalize grassland, efforts are needed to offer space to bats, too, since these animals are important members of the ecosystem—e.g., by reducing the number of pest insects.

And at last, it should be mentioned that the studies on the saliva of the vampire bat *Desmodus rotundus* led to a new and effective enzymatic therapy of aplexia by suspending blood clots (Schleunig 1993; Reddrop et al. 2005; Steiner et al. 2007).

Today bats are protected completely or in part in several countries. In Europe and Russia, all species are protected by law. More and more, the beneficial role of these animals has been recognized in the public. This resulted in increasing efforts to help bats by gates open for bats in caves and buildings, as well as providing special bat houses as roosting sites (Wund and Myers 2005).

References

Arellano-Sota C (1988) Vampire bat-transmitted rabies in cattle. Rev Infect Dis 10(Suppl 4):707–709

Australian Bat Lyssavirus, a Queensland Health fact sheet. http://www.health.qld.gov.au, 18 Mar 2013

Bell GP (1980) A possible case of interspecific transmission of rabies in insectivorous bats. J Mammal 61:528–530

Belotto A, Leanes LF, Schneider MC, Tamayo H, Correa E (2005) Overview of rabies in the Americas. Virus Res 111:5–12

Brass DA (1994) Rabies in bats. Livia Press, Ridgefield, CT

Callisher CH, Childs JE, Fild HE, Holmes VE, Schountz T (2006) Bats: important reservoir of emerging viruses. Clin Microbiol Rev 19:531–545

Carroll RL (1988) Vertebrate paleontology and evolution. W. Freeman and Co., New York, NY

Cockrum EL (1969) Migration in the guano bat, *Tadarida brasiliensis*, vol 51. Miscellaneous Publications, Museum of Natural History, University of Kansas, pp 303–336

Constantine DG (1962) Rabies transmission by nonbite route. Public Health Rep 77:287–289

Constantine DG (1967) Activity patterns of the Mexican free-tailed bat. University of New Mexico publications in biology, vol 7. University of New Mexico Press, Albuquerque, pp 1–79

Fenton MB (1992) Bats. Facts On File, Academic, New York, NY

Francis JR, Mattke AC, Vaska V, McCall B, Hutchinson P, Powell J, Nourse C (2013) Australian bat lyssavirus infection in a child: clinical and public health aspects of a rare and devastating infection. CDC conference 2013, ASID annual scientific meeting 2013, 95

Fujita MS, Tuttle MD (1991) Flying foxes (Chiroptera: Pteropodidae): threatened animals of key ecological and economic importance. Conserv Biol 5:455–463

Geoffroy E (1810) Sur les Phyllostomes et les Megadermes. Ann Mus Hist Nat 15:157–198

Gracheva EO et al (2011) Ganglion-specific splicing of TRPV1 underlies infrared sensation in vampire bats. Nature 476:88–92. doi:10.1038/nature10245

Grzimek B (1975) Grzimek's animal life encyclopedia. Van Nostrand Reinhold Co., New York, NY

Harenberg JC (1733) Vernünftige und Christliche Gedancken uber die Vampirs oder bluhtsaugende Todten, so unter den Türcken und auf den Gräntzen des Servien-Landes lebenden Menschen und Viehe das Bluht aussagen sollen, Begleitet mit allerley theologischen, philosophischen und historischen aus dem Reiche der Geister hergeholten Anmerckungen. Wolffenbüttel 1733. Zu finden bey J.C. Meißner

Hendra Virus Infection, a Queensland Health fact sheet. http://www.health.qld.gov.au, Version: 26, 27 Sept 2012

Hill JE, Smith JD (1984) Bats: a natural history. University of Texas Press, Austin, TX

http://www.ucmp.berkeley.edu/mammal/eutheria/chirolh.html

Jones JK, Genoway HH (1970) Chiroptera systematics. In: Slaughter RH, Walton DW (eds) About bats: a chiropteran symposium. Southern Methodist University Press, Dallas, TX, pp 3–21

Lawlor TE (1979) Handbook of the orders and families of living mammals. Mad River Press Inc., subsequent edition, 330 pp

Lyman CP (1970) Thermoregulation and metabolism in bats, p. 301-330. In: Wimsatt WA (ed) Biology of bats. Academic, New York, NY

Maywald A, Pott B (1988) Fledermäuse: Leben, Gefährdung, Schutz. Ravensburger Buchverlag Ott Maier GmbH

MMWR (2011) Human rabies from exposure to a vampire bat in Mexico—Louisiana, 2010. Centers for Disease Control and Prevention, Morbidity and Mortality weekly report 60/No. 31

Neumeister B, Geiss KH, Braun RW, Kimmig P (eds) (2009) Microbiological diagnostics, 2nd edn. Thieme, Stuttgart

Nowak R (1991) Order Chiroptera in Walker's mammals of the World, vol 1, 5th edn. John Hopkins University Press, Baltimore, MD

Pettigrew JD, Jamieson BGM, Robson SK, Hall LS, McAnally KI, Cooper HM (1989) Phylogenetic relations between microbats, megabats and primates (Mammalia: Chiroptera and Primates). Philos Trans R Soc Lond B Biol Sci 325:489–559

Rabies; a Queensland Health fact sheet. http://www.health.qld.gov.au, Version: 6. 5 Dec 2012

Reddrop C et al (2005) Vampire bat salivary plasminogen activator (desmoteplase) inhibits tissue-type plasminogen activator-induced potentiation of excitotoxic injury. Stroke 36:1241–1246

Schleunig WD (1993) Blutgerinsel auflösender Wirkstoff aus dem Speichel von Vampirfledermäusen. Spektrum der Wissenschaft, Weinheim

Schneider MC, Romijn PC, Uieda W, Tamayo H, Fernandes da Silva D, Belotto A, Barbosa da Silva J, Leanes LF (2009) Rabies transmitted by vampire bats to humans: an emerging zoonotic disease in Latin America? Pan Am Salud Publica/Pan Am J Public Health 25:260–269

Simmons NB (2005) Order Chiroptera. In: Wilson DE, Reeder DM (eds) Mammal species of the world: a taxonomic and geographic reference, 3rd edn. Johns Hopkins University Press, Baltimore, MD

Simmons NB, Conway T (1997) Chiroptera. Bats. http://tolweb.org/Chiroptera/15966/1997.01.01 in The Tree of Life web project, Version 1 Jan 1997

Simmons NB, Seymour KL, Habersetzer J, Gunnell GF (2008) Primitive early Eocene bat from Wyoming and the evolution of flight and echolocation. Nature 451:818–821. doi:10.1038/nature06549

Smith JD (1976) Chiropteran evolution. In: Baker RJ, Jones JK, Carter DC (eds) Biology of bat of the New World family Phyllostomatidae, part I. Texas Tech University, Lubbock, TX, pp 49–69, Special Publications of the Museum, no. 10

Steiner T, Jüttler E, Ringleb P (2007) Akuttherapie des Schlaganfalls. Der Nervenarzt 78(10): 1147–1154

Tuttle MD (1990) America's neighborhood bats. Texas Press, Austin, TX, pp 17–21

Vampire Bats. Smithsonian Information, http://www.si.edu/Encyclopedia_SI/nmnh/batfacts.htm, Nov 1980

Westheide W, Rieger R (eds) (2010) Spezielle Zoologie, Teil 2: Wirbel- und Schädeltiere, 2nd edn. Spektrum Akademischer Verlag, Heidelberg

Wilkinson GS (1990) Food sharing in vampire bats. Sci Am 76–82

Wimsatt WA (1970) Preface, p.xi–xii. In: Wimsatt WA (ed) Biology of bats. Academic, New York, NY

Winkler WG (1968) Airborn rabies virus isolation. Bull Wildl Dis Assoc 4:37–40

Wund M, Myers P (2005) "Chiroptera" (On-line). Animal diversity web. Accessed 27 Apr 2013 at http://animaldiversity.ummz.umich.edu/accounts/Chiroptera/

Chapter 3
Bats as Potential Reservoir Hosts for Vector-Borne Diseases

Christian Melaun, Antje Werblow, Markus Wilhelm Busch, Andrew Liston, and Sven Klimpel

Abstract Bats are the only mammals with the capacity of powered flight. Nearly 1,000 species can be found all over the world except in the northern and southern polar areas. They perform important ecosystem services such as control of insects, reseeding of cut forests and pollination of plants, which provide food for humans and animals. On the other side, they are also recognized to be natural reservoir hosts of a large variety of zoonotic diseases with the ability to cross species barriers. To date, more than 80 virus species of different groups and various parasites, which can cause several diseases have been isolated or detected in bats. Especially their high population density and gregarious roosting behaviour increase the likelihood of intra- and inter-species transmission of infections. Another important factor, which enables pathogens to spread long distances, is the migratory habit of some bat species, resulting in a great dispersal capacity. The transmission of pathogens from bats to humans or other animals occurs by direct contact with infected animals, their blood and tissue or through vector species. One of the most important vector groups are insects. With more than a million described species, they are the most diverse group of animals. Especially haematophagous groups such as Cimicidae, Culicidae or Phlebotominae are known as vectors for a variety of diseases. These include bacteria, protozoan and metazoan parasites as well as viruses. We focused on blood-feeding insects, because the presence of certain viruses in them as well as in bats comprises a potential virus transmission from bats to humans through mosquitoes or other blood-feeding insects. For this chapter, we could find 20 viruses

---

C. Melaun (✉) • A. Werblow • M.W. Busch • S. Klimpel (✉)
Biodiversity and Climate Research Centre (BiK-F), Medical Biodiversity and Parasitology; Senckenberg Gesellschaft für Naturforschung (SGN); Goethe-University (GO), Institute for Ecology, Evolution and Diversity, Senckenberganlage 25, 60325 Frankfurt am Main, Germany
e-mail: Christian.Melaun@senckenberg.de; klimpel@bio.uni-frankfurt.de

A. Liston
Senckenberg German Entomological Institute (SDEI), Eberswalder Str. 90, 15374 Müncheberg, Germany

S. Klimpel and H. Mehlhorn (eds.), *Bats (Chiroptera) as Vectors of Diseases and Parasites*, Parasitology Research Monographs 5, DOI 10.1007/978-3-642-39333-4_3,

from four different families and two parasitic pathogens detected in all three groups of haematophagous insects.

Keywords Chiroptera • Haematophagous insects • Vectors • Virus • Diseases • Parasites

3.1 Introduction

Bats (order Chiroptera) are one of the most diverse, abundant and widely distributed groups of mammals and the only one with the capacity of powered flight (Li et al. 2010; Omatsu et al. 2007). Nearly 1,000 species are found worldwide, except in the northern and southern polar areas, representing approximately 20 % of all mammalian species (Omatsu et al. 2007; Teeling et al. 2005). Chiroptera can be divided into the two suborders Megachiroptera (old world fruit bats) and Microchiroptera (echolocating bats) (Jones et al. 2002). Analysis of 17 nuclear genes dated the origin of chiropterans up to 50 million years ago (Calisher et al. 2006). One major lineage of Microchiroptera was traced to Laurasia and one to Gondwana (Teeling et al. 2005). Derived ancient origins for certain zoonotic viruses in bats, such as lyssa and henipa viruses, suggest a long history of coevolution and cospeciation (Calisher et al. 2006). On the one hand, bats perform important ecosystem services, such as control of insects (Reiskind and Wund 2009, 2010; Rydell et al. 2002), reseeding of cut forests and pollination of plants that provide food for humans and animals. On the other hand, bats are recognized to be a natural reservoir of a large variety of zoonotic viruses, which can cross species barriers to infect humans and other domestic or wild animals (Li et al. 2010). To date, more than 80 virus species of different groups have been isolated or detected in bats. Bats have a great dispersal capacity and the migratory habits of some species provide a good opportunity for pathogens to spread long distances (Messenger et al. 2003). Further, different migration patterns can often be found within one species. These different patterns may allow the exchange of viruses or virus variants between subpopulations as well as members of other species, because several species may roost in the same place. Also, their high population densities and gregarious roosting behaviour increase the likelihood of intra- and inter-species transmission of viral infections (Calisher et al. 2006). The extreme longevity of bats may help to maintain the viruses, resulting in chronic infections, and increase the chances of transmission to other mammals or vertebrates. Calisher et al. (2006) suggest an explanation why some viruses, which are deadly for humans and other mammals, can persist in bats without proving being fatal: because bats form a very ancient lineage amongst the mammals, it is possible that their immune system with its innate acquired immune responses may differ significantly from those of other mammals. So far very little is known about bat immune systems, although some studies indicate similarities between bats immune responses and those of other mammals (Chakravarty and Sarkar 1994; McMurray et al. 1982; Sarkar and

Chakravarty 1991). The studies of Halpin et al. (2000), Lau et al. (2005) and Leroy et al. (2005) show the occurrence of virus-specific B- and T-cell responses despite persistent virus infection. One possible pathway of virus transmission involves the animal's ecology. Due to their flying habits, bats are constrained by the aerodynamics of flight and cannot therefore ingest huge amounts of food. Instead of swallowing whole fruits, bats chew these to extract sugars and other substances. The partially digested fruit is spat out and falls to the ground, where these remnants are fed on by other animals. Residual virus particles in the bat saliva on the fruit remnants may cause infection of the latter animals. Heavier body parts of insect prey are discarded in the same way and are also eaten by terrestrial foraging species (Dobson 2005).

Insects comprise the most diverse group of animals with more than a million described species. The estimated number of extant species is between four and six million (Chapman 2009; Novotny et al. 2002). Insects can be found in nearly all terrestrial environments. Some species have become specialized feeders on blood. Depending on the species, this haematophagous behaviour can be observed either in both males and females or just in females. In the latter case, blood proteins are essential for egg production. All haematophagous insects use modified extremities of the head and extensions of the head capsule as piercing-sucking mouthparts to obtain and feed on blood (Krenn and Aspöck 2012; Lehane 2005). Convergent evolution has led to the development of piercing proboscides in various haematophagous insect groups (Krenn and Aspöck 2010, 2012).

One group of haematophagous insects with a worldwide distribution is the family Cimicidae (bed and bat bugs), which contains more than 100 species. Balvín (2008) mentions that bats are generally considered to be the original hosts of the family. The same had long been thought about the bedbug *Cimex lectularius* (e.g. Sailer 1952; Usinger 1966), but new results suggest an early sympatric speciation on humans and bats for different populations of bed bugs (Balvín et al. 2012). Although various pathogens have been identified from Cimicidae (e.g. Burton 1963; Delaunay et al. 2011; Goddard and deShazo 2009), their role as vectors is still unclear.

Mosquitoes (Diptera: Culicidae) are a second group of blood-sucking insects. They are regarded worldwide as the major vectors of vector-borne diseases. Especially, species of the genus *Anopheles*, which are well known as vectors of malaria, can also impact human and animal health by their ability to transmit arboviruses (arthropod-borne viruses) as well as filarial parasites such as the elephantiasis-causing *Wuchereria bancrofti* and *Brugia malayi* (Gillies and Coetzee 1987; Sallum et al. 2000; Service 1993). Apart from *Anopheles* spp., numerous other species of mosquitoes are pests or vectors of pathogens that cause diseases in humans and animals (Reinert et al. 2004). Because of their adaptive abilities, mosquitoes are capable of inhabiting and surviving in a wide range of habitats. Worldwide, they colonize nearly every aquatic habitat. As a result of their large flight range, some floodwater mosquito species can become pests even in places located far away from their breeding sites (Mohrig 1969; Schäfer et al. 1997). Additionally, flood plains along coastal areas as well as tree holes are used as breeding sites by certain species. Also impressive is the adaptive capacity of mosquitoes to extremes of, or changes in, climatic factors. This ecological flexibility is one of the reasons for the success of mosquitoes (Becker et al. 2010). With the exception of the tropical genus *Toxorhynchites*, in which both sexes subsist on carbohydrate-rich materials such as honeydew, nectar and plant fluids (Snodgrass 1959), the females consume blood, e.g. to obtain proteins necessary for egg production. The host is located mainly by their olfactory senses, by the odour of carbon dioxide or the use of visual contact (Becker et al. 2010). While some mosquito species have specialized on birds, amphibians and other animal groups, others feed on mammals. Especially species with the last feeding habits, or possibly hybrids between different species, can be responsible for zoonoses, because they act as vectors between reservoir hosts and humans. Phlebotomine sand flies (Diptera: Psychodidae: Phlebotominae) are small (ca. 3 mm) nematoceran dipterans (Ready 2013) and a third group of haematophagous insects. They are mainly distributed in the tropics, but there is also an important Palaearctic element (Lewis 1974). The group acts as vector of leishmaniasis, *Bartonella bacilliformis* as well as some arboviruses of the three different genera: *Phlebovirus* (family Bunyaviridae), *Vesiculovirus* (family Rhabdoviridae) and *Orbivirus* (family Reoviridae) (Depaquit et al. 2010). Known disease agents, who can be transmitted by insects and have a potential impact on bats include bacteria, viruses, fungi as well as protozoan and metazoan parasites. With a few exceptions, in this chapter we concentrate on viruses, which have been detected in bats, haematophagous insects and humans (short information see Table 3.1).

3.2 Pathogens in Bats, Humans and Haematophagic Insects

The most important pathogens causing viral infections in bats, humans and haematophagic insects belong to the families Togaviridae (genera *Alphavirus* and *Rubivirus*), Flaviviridae (genus *Flavivirus*), Bunyaviridae (genera *Orthobunyavirus* and *Phlebovirus*) and Arenaviridae (genus *Arenavirus*).

3.2.1 Bunyaviridae

The Bunyaviridae is the largest family of RNA viruses with over 300 serologically or molecular-genetically distinguishable strains (Elliott 1997; Soldan and González-Scarano 2005). It was characterized in 1975 (Soldan and González-Scarano 2005) and is now considered to contain five genera: *Tospovirus* (the only plant-infecting viruses in the group), *Hantavirus*, *Nairovirus*, *Phlebovirus* and *Orthobunyavirus* (Bowen et al. 2001; LeDuc and Kahlon 2012; Weidmann et al. 2003). The medically most important pathogens within this family are not only transmitted through the bite of infected mosquitoes but also by sand flies or

Table 3.1 Overview of the diseases mentioned in this review, further details incl. sources can be found in the particular chapters (AB = Antibodies, n.k. = not known)

virus family	genus	virus species	clinical features	disease occurence	bat species	common name	distribution infect. bat	vector	*vector species*
Arenaviridae	Arenavirus	Tacaribe virus (TACV)	haemorrhagic fever	South America	*Artibeus lituratus* *Artibeus jamaicensis* *Desmodus rotundus* (AB) *Platyrrhinus helleri* (AB) *Sturnira lilium* (AB)	Great fruit eating bat Jamaican fruit bat Common vampire bat Heller's broad-nosed bat Yellow-shouldered bat	Trinidad/Guatemala Trinidad n.k. n.k. n.k.	mosquito	n.k.
Bunyaviridae	Orthobunyavirus	Bwamba virus (BWAV)	fever headache myalgia slow pulse	Cameroon Kenya Nigeria Uganda	*Myotis lucifugus*	Little brown bat	infection only in the Laboratory	mosquito	*Aedes furcifer* *Anopheles coustai* *Anopheles funestus* *Anopheles gambiae* *Mansonia uniformis*
		Catú-virus (CATUV)	fever headache body pains weakness	Brazil	*Mollosus currentium*	Bonda mastiff bat	Brazil	mosquito	*Anopheles nimbus* *Culex* spp.
		Guamá-virus (GMAV)	fever headache muscle pain myalgia leukopenia	Brazil	unidentified	n.k.	Brazil	mosquito sandfly	*Culex* spp. n.k.
		Kaeng Khoi virus (KKV)	fibrile illness	Cambodia Thailand	*Chaerephon plicatus* *Taphozous theobaldi*	Wrinkle-lipped free-tailed bat Theobald's tomb bat	Thailand Cambodia	bedbugs	*Stricticimex parvus* *Cimex insuetus*
	Phlebovirus	Rift Valley fever virus (RVFV)	body pain flu-like fever loss of memory neck stiffness retinal lesions sensitivity to light vomitting	Egypt Kenya Madagascar Mauritania Saudi Arabia Somalia Yemen	*Epomops franqueti* *Glauconycteris argentata* *Hipposideros abae* *Hipposideros caffer* *Micropteropus pusillus* *Miniopterus schreibersii*	Franquet's epauletted fruit bat Silvered bat Aba roundleaf bat Sundevall's roundleaf bat Peter's dwarf epauletted fruit bat Schreiber's long-fingered bat	Guinea n.k. Guinea n.k. n.k. n.k.	mosquito sandfly	*Aedes dalzieli* *Aedes dentatus* *Aedes circumluteolus* *Culex antennatus* *Culex neavei* *Culex pipiens* *Phlebotomus duboscqi* *Phlebotomus papatasi*
		Toscana Virus (TOSV)	aseptic meningitis headache meningoencephalitis muscle aches neck stiffness strong fever	France Greece Italy Spain Sweden Swiss	*Pipistrellus kuhlii*	Kuhl's pipistrelle	n.k.	sandfly	*Phlebotomus perniciosus* *Phlebotomus perfiliewi*
Flaviviridae	Flavivirus	Dengue virus (DENV)	general indisposition headache haemorrhagic fever orbital pain	tropical and subtropical regions, especially in southeast and south Asia	*Artibeus jamaicensis* (AB) *Carollia perspicillata* *Molossus* sp. (AB) *Myotis nigricans* (AB) *Natalus stramineus* (AB) *Pteronotus parnellii* (AB) *Uroderma* sp.(AB)	Jamaican fruit-eating bat Seba's short-tailed bat n.k. Black myotis Mexican funnel-eared bat Parnell's mustached bat n.k.	Mexico French Guinea n.k. Mexico Mexico Mexico n.k.	mosquito	*Aedes aegypti* *Aedes albopictus*
		Ilheus virus (ILHV)	chills CNS pleocytosis fever photophobia asthenia	Bolivia Ecuador Panama Trinidad	unidentified	n.k.	n.k.	mosquito	*Culex* spp. *Haemagogus spp.* *Ochlerotatus spp.* *Psorophora ferox* *Sabethes spp.* *Trichoprospon spp.*

(continued)

Table 3.1 (continued)

virus family	genus	virus species	clinical features	disease occurence	bat species	common name	distribution infect. bat	vector	*vector species*
Flaviviridae	Flavivirus	Japanese encephalitis virus (JEV)	CNS pleocytosis CNS signs fever headache prostration	Asia northern Australia	*Miniopterus schreibersi* *Myotis mystacinus* (AB) *Plecotus auritus* (AB) *Pipistrellus abramus* (AB) *Rhinolophus pusillus*	Schreiber's long-fingered bat Whiskered bat Brown big-eared bat Japanese Pipistrelle Least horseshoe bat	Japan n.k. n.k. n.k. Japan	mosquito	*Culex annulirostris* *Culex gelidus* *Culex pipiens* *Mansonia uniformis* *Ochleorotatus japonicus*
		St. Louis encephalitis (SLEV)	fever headache meningoencephalitis	Canada Central/South America USA	*Eptesicus fuscus* (AB) *Myotis lucifugus* (AB)	Big brown bat Little brown bat	USA USA	mosquito	*Aedes aegypti* *Culex nigripalpus* *Culex quinquefasciatus* *Culex tarsalis*
		West Nile virus (WNV)	coma fever headache neck stiffness paralysis tiredness	Africa Asia Middle East Europe	*Eptesicus fuscus* (AB) *Myotis lucifugus* (AB) *Myotis septentrionalis* (AB) *Rousettus aegyptiacus* *Rousettus leschenaultii* *Tadarida brasiliensis* (AB)	Big brown bat Little brown bat Northern myotis Fruit bat Leschenault's rousette Brazilian free-tailed bat	India New York n.k. Uganda/Israel India n.k.	mosquito	*Coquillettidia richiardii* *Culex modestus* *Culex pipiens* *Culex quinquefasciatus* *Culex univittatus* *Culex vishnui*
		Yellow fever virus (YFV)	body aches chills fever headache nausea vomiting	Africa Latin America	*Eidolon* sp. *Rousettus* sp.	n.k. n.k.	East Africa	mosquito	*Aedes* spp. *Haemogogus spp.*
		Zika virus (ZIKV)	diarrhea dizziness fever headache joint pain rash	Gaboon Sierra Leone Tanzania Thailand Uganda Yap Island	*Myotis lucifugus*	Little brown bat	infection only in the Laboratory	mosquito	*Aedes aegypti* *Aedes africanus* *Aedes apicoargenteus* *Aedes furcifer* *Aedes luteocephalus* *Aedes vittatus*
Togaviridae	Alphavirus	Chikungunya virus (CHIK)	fever headache haemorrhagic signs leukopenia lyphadenopathy	Africa south and southeast Asia	*Chaerephon pumilus* *Hipposideros caffer* *Rousettus leschenaultia* *Scotophilus* sp.	Little free-tailed bat Sundevall's leaf-nosed bat Leschenault's rousette Asiatic yellow bat	China China China Senegal	mosquito	*Aedes aegypti* *Aedes albopictus*
		Eastern equine encephalitis virus (EEV)	abdominal pain fever cyanosis lethargy seizures	Eastern USA	*Artibeus lituratus* (AB) *Artibeus jamaicensis* (AB) *Eptesicus fuscus* *Lasiurus cinereus* *Myotis lucifugus* *Tadarida brasiliensis* (AB)	Great fruit-eating bat Jamaican fruit-eating bat Big brown bat Hawaiian hoary bat Little brown bat Brazilian free-tailed bat	Guatemala Guatemala New Jersey New Jersey New Jersey n.k.	mosquito	*Aedes vexans* *Coquillettidia perturbans* *Culiseta melanura* *Culex erraticus* *Uranotaenia sapphirina*
		Ross River virus (RRV)	arthralgia/arthritis fever paraesthesia rash	Am. Samoa Australia Fiji New Guinea Solomon Is. S. Pacific Is	*Pteropus conspicillatus*	Spectacled flying-fox	Australia	mosquito	*Aedes camptorhynchus* *Aedes vigilax* *Culex annulirostris*
		Sindbis virus (SINV)	headache itchy exanthema joint pain mild fever nausea	Africa Asia Australia Europe	*Hipposideridae* sp. *Rhinolophidae* sp.	Old world leaf-nosed bat Horseshoe bat	Venezuela Mexiko	mosquito	*Aedes* spp. *Culiseta morsitans* *Culex pipiens* *Culex torrentium* *Ochleorotatus* spp.

Togaviridae	Alphavirus	Venezuelan equine encephalitis virus (VEEV)	ataxia chills coma fever headache malaise nuchal rigidity paralysis sore throats	Belize Colombia Ecuador El Salvador Guatemala Honduras Mexico Panama Venezuela	*Artibeus lituratus* *Artibeus jamaicensis* (AB) *Artibeus phaeotis* (AB) *Desmodus rotundus* (AB) *Glossophaga commissarisi* (AB) *Myotis nigricans* (AB) *Uroderma bilobatum* (AB)	Great fruit-eating bat Jamaican fruit-eating bat Pygmy fruit-eating bat Common vampire bat Commissaris's long-tongued bat Black myotis Tent-making bat	Guatemala	mosquito	*Aedes serratus* *Culex* spp. *Mansonia titilans* *Ochleorotatus taeniorhynchus*
		Western equine encephalitis (WEEV)	fever headache manifestations in CNS somnolence	North America	*Artibeus lituratus* (AB) *Artibeus jamaicensis* (AB)	Great fruit-eating bat Jamaican fruit-eating bat	Guatemala Haiti	mosquito	*Aedes campestris* *Aedes dorsalis* *Culex tarsalis* *Ochleorotatus melanimon*

disease	parasite	species	symptoms	occurence	bat species	common name	distribution bat	vector	*vector species*
Chagas-disease	Flagellates	*Trypanosoma cruzi*	fever swelling around the site of inoculation inflammation of heart, muscle or brain	Central and South America	*Artibeus lituratus* *Carollia perspicillata* *Desmodus rotundus* *Glossophaga soricina* *Molossus molossus* *Phyllostomus hastatus*	Great fruit-eating bat Seba's short-tailed bat Common vampire bat Pallas's long-tongued bat Pallas's mastiff bat Greater spear-nosed bat	infection under controlled conditions, but also natural	triatomine bug	*Rhodnius prolixus* *Triatoma dimidiata* *Triatoma infestans*
Malaria	Apicomplexa	*Plasmodium* sp. *Hepatocytis* sp. *Polychromophilus* sp.	fever headache death	Africa Mediterranean South-East Asia	*Eptesicus serotinus* *Kerivoula hardwickii* *Megaderma spasma* *Miniopterus gleni* *Myotis daubentonii* *Myotis goudoti* *Myotis myotis* *Nyctalus noctula* *Pteropus poliocephalus*	Serotine Hardwicke's woolly bat Lesser false vampire Glen's long-fingered bat Daubenton's myotis Malagasy mouse-eared bat Greater mouse-eared bat Noctule Grey-headed flying fox	Switzerland Cambodia Cambodia Madagascar Switzerland Madagascar Switzerland Switzerland Australia	mosquito biting-midges bat flies	*Anopheles* spp. *Culicoides nubecolusus* *Nycteribia kolenatii*

possibly by bedbugs (Darai et al. 2011; Elliott and Blakqori 2011). The largest of the five genera is the genus *Orthobunyavirus*, containing 174 known viruses (Elliott and Blakqori 2011). Within this group, up to half of the potential 60 Bunyaviridae viruses that cause disease in humans belong to *Orthobunyavirus* (Soldan and González-Scarano 2005) including the prototype bunyavirus, Bunyamwera virus. It was first isolated in 1943 from *Aedes* mosquitoes in Uganda and gave its name to the family Bunyaviridae and the genus *Bunyavirus*. In 2005, Bunyavirus was renamed *Orthobunyavirus* (Bowen et al. 2001; Elliott and Blakqori 2011). The first virus within this genus is the **Bwamba virus** (BWAV), which is a member of the Bwamba serogroup (Lambert and Lanciotti 2008). It is transmitted by mosquitoes including *Aedes furcifer, Anopheles coustani, Anopheles funestus, Anopheles gambiae* and *Mansonia uniformis* (Lee et al. 1974; Lutwama et al. 2002). According to Gonzales and Georges (1988), the principal anthropophilic vector species are *An. funestus* and *An. gambiae*. The first encounter with Bwamba fever was in 1937 among construction workers in Western Uganda (Smithburn et al. 1941). Today, Bwamba virus is endemic in Nigeria, Cameroon, Central African Republic, Kenya, Tanzania and South Africa, but due to the mild symptoms it is often mistaken for malaria (Lutwama et al. 2002; Moore et al. 1975; Smithburn et al. 1941). In a laboratory study by Reagan et al. (1955), the cave bat *Myotis lucifugus* was successfully infected after intraperitoneal, intradermal, intracerebral and intrarectal injection of the virus, although the bats were not susceptible to the virus after intranasal exposure. The second virus is the **Kaeng Khoi virus** (KKV), which was first isolated in Thailand in 1969 from the bat species *Chaerephon plicata* (wrinkle-lipped bat) and *Taphozous theobaldi* (Theobald's bat). Both species can be found across the Indian subcontinent and Southeast Asia. (Hutson et al. 2001) In 1976 and 2003, the virus was found again in *C. plicata* in Thailand and for the first time in Cambodia (Osborne et al. 2003; Williams et al. 1976). Apart from bats, Williams et al. (1976) found the virus also in bedbugs (*Stricticimex parvus* and *Cimex insuetus*) that inhabit caves together with other haematophagous arthropods, which attack humans. The virus might be a public health concern, because serum analysis found neutralizing antibody in 29 % of the population. The symptoms of the virus in bats and humans are unknown, but a survey of the population indicated that they believe that bedbug bites were the cause of an influenza-like illness, which is typical of infection by members of *Orthobunyavirus* (Osborne et al. 2003). Also known to occur in bats and humans are the **Guamá and Catú viruses**, which are members of the Guamá serotype group and are transmitted by species of *Culex* mosquitoes (Darai et al. 2011; Löscher and Burchard 2008). Both were isolated from humans and mosquitoes in the Amazon area (Causey et al. 1961) and Catú virus also from humans in Trinidad (Tikasingh et al. 1974). However, Catú virus was also isolated in Brazil from the bat *Molossus currentium* and Guamá virus from an unidentified bat (Calisher et al. 2006; Karabatsos 1985). The symptoms of both viruses show considerable variation such as fever, headache, general body pains, weakness or dizziness and photophobia (Causey et al. 1961).

Another group of viruses that may be transmitted by bats is the genus ***Phlebovirus***, including nine species with 37 viruses (Bouloy 2011), distributed in

Africa, Asia, North and South America and the Mediterranean region (McMullan et al. 2012). Many phleboviruses are transmitted by sandflies or other arthropods such as mosquitoes or ticks. The first pathogen in this genus that has to be considered in connection with bats is the **Rift Valley fever virus** (RVFV). It is primarily transmitted by mosquitoes of the genus *Aedes* (e.g. *Ae. cumminsii, Ae. circumluteolus, Ae. mcintoshi* or *Ae. vexans*) or *Culex* (e.g. *Cx. pipiens, Cx. tritaeniorhynchus* or *Cx. neavei*), but it has been shown that sandflies (*Phlebotomus duboscqi* and *P. papatasi*) might also be potential vectors (Dohm et al. 2000; Fontenille et al. 1998; Pepin et al. 2010). RVFV can be transmitted into mosquito offspring transovarially (Ikegami and Makino 2011). Outbreaks of RVF are associated with heavy rainfalls during the El Niño/Southern Oscillation (ENSO) phenomenon (Miller et al. 2002; WHO 2013a) because the floods create optimal breeding conditions (Bowen et al. 2001; Fontenille et al. 1998; Woods et al. 2002). After the first identification of the virus in the Rift Valley of Kenya in 1930, numerous outbreaks of RVF have been reported in many regions of Africa (Fig. 3.1a) (Daubney and Hudson 1931; Fontenille et al. 1998). The largest occurred in Egypt in 1977–1978 with 200,000 estimated human infections, 18,000 cases of illness and 600 deaths (Ikegami 2012). The first detected outbreak of RVF outside the African continent was in 2000 in Saudi Arabia and Yemen. The latest outbreak of RVF was 2012 in Mauritania with 34 cases and 17 deaths reported (WHO 2012). The clinical symptoms range from flu-like fever, muscle pain or headache to neck stiffness, retinal lesions, loss of memory and even death (Ikegami 2012; WHO 2013b). However, RVF occurs not only in humans. Outbreaks can also result in devastating economic losses when livestock is infected (Woods et al. 2002). In 1991, the virus was also isolated from bats in West Africa (Fontenille et al. 1998). Calisher et al. (2006) suggest the bat species *Micropteropus pusillus, Epomops franqueti, Hipposideros abae, H. caffer, Miniopterus schreibersii* and *Glauconycteris argentata* as potential hosts for the RVFV. They are all distributed in West, East and Central Africa except for *M. schreibersii*, which occurs primarily in southern and middle Europe (e.g. Portugal, Italy (Sardinia, Sicily), Turkey, Switzerland and Serbia) (IUCN red list 2012). The effect on potential vectors (e.g. *Aedes vexans*) of climate changes may enable this bat to spread the virus to European countries. The second pathogen within the *Phlebovirus* genus that has been linked with bats is the **Toscana virus**, a serotype of Sand fly fever distributed in the Mediterranean region (Fig. 3.1b) (Charrel et al. 2005; Cusi et al. 2010). In 1971, the Toscana virus was isolated for the first time from the sand fly *Phlebotomus perniciosus* in Monte Argentario, Toscana (Braito et al. 1997; Cusi et al. 2010; Valassina et al. 2003; WHO 2004). Later the virus was also isolated from the sand fly *P. perfiliewi* (Valassina et al. 2003). However, *P. perniciosus* is probably the most common insect vector of the Toscana virus, being one of the most abundant sand flies in southern Europe (Maroli et al. 1994; Sanbonmatsu-Gámez et al. 2005). With tourists, returning from Mediterranean countries, the virus has been imported to further European countries such as Germany, Sweden and Switzerland (Cusi et al. 2010). The only known vertebrate hosts beside humans are bats. Accordingly, the virus has been isolated from the bat

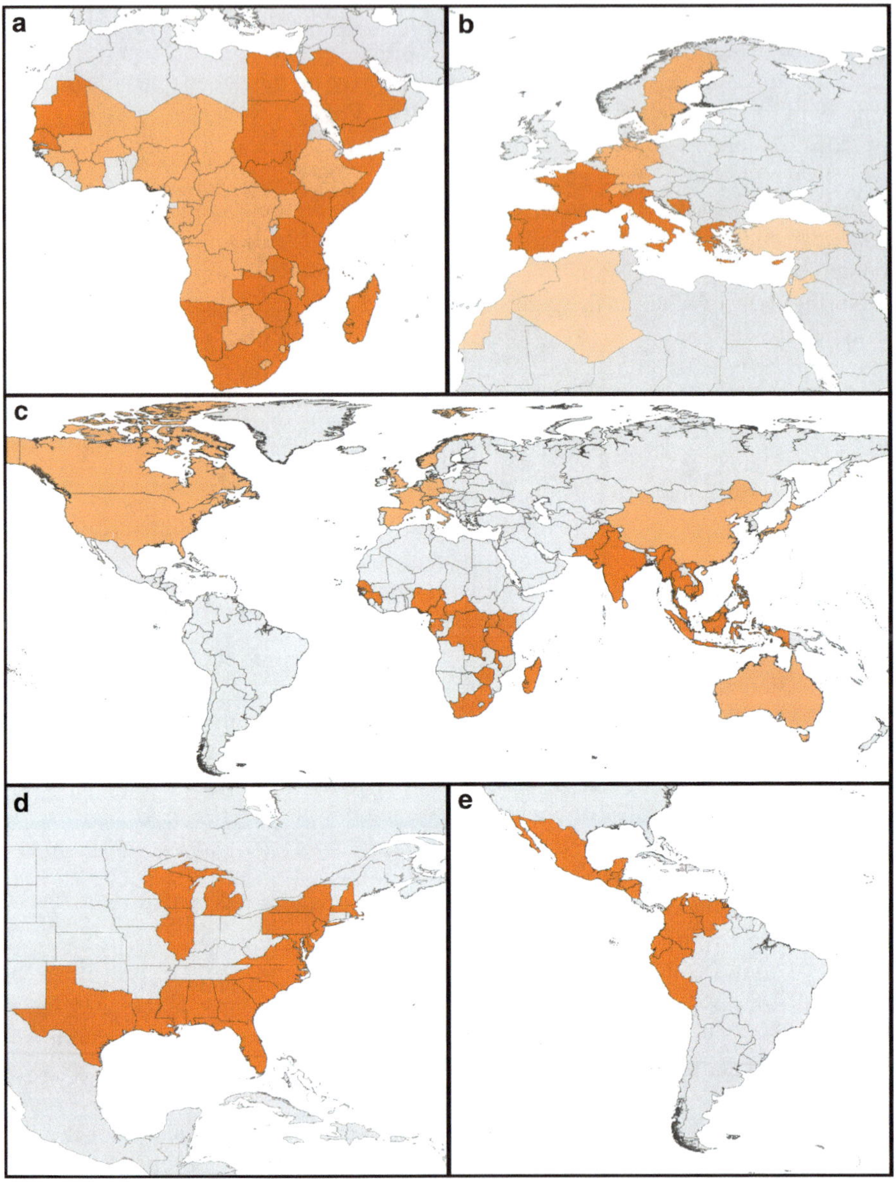

Fig. 3.1 Virus distribution (**a**) Rift Valley fever: *dark orange* (do)—epizootic and epidemic areas; *light orange* (lo)—serological evidence or virus isolation (Ikegami 2012). (**b**) Toscana virus: do—native infections; *middle orange* (mo)—imported cases; lo—seropositive cases in population (Cusi et al. 2010). (**c**) Chikungunya virus: do—endemic or epidemic areas; lo—imported cases (Powers and Logue 2007). (**d**) Eastern equine encephalitis: reported cases between 1964 and 2010 (CDC 2010). (**e**) Venezuelan equine encephalitis: major outbreaks regions (Weaver et al. 2004)

Pipistrellus kuhlii in regions where the insect vectors were present (Charrel et al. 2005; Valassina et al. 2003). The role of bats as virus reservoirs is not yet definitely clarified (Valassina et al. 2003). Most infections are recorded during summertime and cause high fever, headaches, muscle aches, neck stiffness and aseptic meningitis with a non-fatal course or a mild meningoencephalitis, but asymptomatic infections are also possible (Baldelli et al. 2004; Braito et al. 1997; Hemmersbach-Miller et al. 2004). Studies in central Italy showed that 52 % of aseptic meningitis cases were caused by Toscana virus (Charrel et al. 2005). Studies of Sanbonmatsu-Gámez et al. (2005) in Spain, where 979 persons were probed for their seroprevalence, indicated that Toscana virus occurs more often in rural areas (26.7 %) than in urban (20.6 %).

3.2.2 Togaviridae

Within this second large virus family, the two genera *Alphavirus* and *Rubivirus* are known (Rolle and Mayr 2007). It can be characterized as a group of positive stranded RNA viruses, which possess a cubically symmetric capsid. All vector-borne togaviruses, which are medically important, belong to *Alphavirus* that contains 26 different viruses (Laine et al. 2004). The first of five viruses associated with bats, human and haematophagous insects is the **Western equine encephalitis virus,** which is eponymous for the western equine encephalitis antigenic complex and an important pathogen not only in human but also in veterinary medicine (Reisler et al. 2012). It was isolated in 1930 from a horse brain although epidemics had already been described in 1912 and 1919 (Johnson 1964). The natural vertebrate hosts of WEEV are passerine and domestic birds (Eklund 1954) where it maintains an enzootic cycle with the mosquito vectors. The main vector is *Culex tarsalis*, of which hybrid strains have evolved that are highly resistant to WEEV infection (Hardy et al. 1978). Other vectors are *Aedes campestris, Ae. dorsalis* and *Ochlerotatus melanimon* (Zacks and Paessler 2010). In early experiments, transmission could also be shown in *Ae. aegypti* (Kelser 1933; Merrill and Ten Broeck 1935; Merrill et al. 1934), *Ae. albopictus* (Simmons et al. 1936), *Ae. nigromaculis* (Madsen and Knowlton 1935; Madsen et al. 1936), *Ae. sollicitans* (Merrill et al. 1934), *Ae. taeniorhynchus* (Kelser 1937, 1938) and *Ae. vexans* (Kelser 1937, 1938; Knowlton and Rowe 1935). *Culiseta* spp. is mentioned as vector by Whitley and Gnann (2002) without mentioning the species or the source of the information.

The connection between Western equine encephalitis and human disease was proven in 1938, when the virus was recovered from a child's brain that died of encephalitis (Griffin 2001). Nevertheless in North America, WEEV is a rare cause of encephalitis and only seven cases were reported from 1987 to 2002 (CDC 2002; Romero and Newland 2003). The virus has caused encephalitis epidemics in emus, humans and horses with fatality rates of 10 % for emus and humans and 20–40 % for horses (Ayers et al. 1994; Nalca et al. 2003). The patients suffer a 2- or 3-day period with fever, headache, vomiting, nausea, somnolence and irritability before manifestations in the CNS begin (Nalca et al. 2003). Constantine (1970) mentions that WEEV has been isolated from bats and Western equine encephalitis N antibodies were detected in *Artibeus jamaicensis* from Haiti (McLean et al. 1979) and *Artibeus lituratus* from Tikal, Guatemala (Ubico and McLean 1995). The latter authors postulate that bats may become infected, especially during epizootics in other hosts. The prototype virus of the genus *Alphavirus* is the **Sindbis virus** (SINV), which belongs together with the Highland J, Fort Morgan, Buggy Creek and Aura virus to the Western equine encephalomyelitis antigenic complex (Hubálek 2008; Lundström and Pfeffer 2010; Netolitzky et al. 2000). It was first recognized and isolated from *Culex pipiens* and *Cx. univittatus* in 1952 in Egypt (Jöst et al. 2010; Kurkela et al. 2008; Laine et al. 2004). The first record of Sindbis virus in European countries was in 1975 (Laine et al. 2004). Today, Sindbis virus is one of the most widely distributed viruses, having been isolated in Europe (e.g. Sweden, Finland and Italy (Sicily)), Africa (e.g. Egypt, Kenya and South Africa), Asia (e.g. China, Malaysia and Lebanon) and Australia (Lundström and Pfeffer 2010; Norder et al. 1996; Tesh 1982). Nucleotide sequencing of SINV strains isolated around the globe has shown that the majority of SINV strains are geographically distinct genotypes and that migrating birds may carry the virus over long distances (Modlmaier et al. 2002; Strauss and Strauss 1994). Together with a wide distribution, the virus also has a broad host range and has been isolated from frogs, ticks, humans and numerous mosquito species (Kurkela et al. 2008; Modlmaier et al. 2002). Furthermore, Blackburn et al. (1982) isolated the virus from the organs of the Microchiropteran bats *Hipposideros* spp. and *Rhinolophus* spp. Vectors of SINV are ornithophilic mosquitoes *Cx. torrentium* or *Cx. pipiens*, *Culiseta morsitans* and *Ochlerotatus* spp. or *Aedes* spp. (Hubálek 2008; Jöst et al. 2010). The first description of symptoms caused by Sindbis virus infection is from Uganda in 1961 (Laine et al. 2004). They range from itchy exanthema, mild fever and joint pain in wrists, hips, knees and ankle to nausea, headache and muscle pain (Kurkela et al. 2005). Together with other mosquito-borne diseases like the Ross River virus, Mayaro-, onyong-nyong-, Bebaru-, Getah and Semliki forest virus, the **Chikungunya virus** is a member of the Semliki forest antigenic complex (Burt et al. 2012). The first isolation of the virus from an infected patient was carried out by Ross in 1952/1953 during an outbreak in Tanzania (Diallo et al. 1999; Tesh 1982; Tiawsirisup 2011). Today, the geographic distribution of this enzootic virus includes the tropical and subtropical areas of Africa, and southern or southeast Asia, including India, Sri Lanka, Myanmar, Thailand, Indonesia, Malaysia and the Philippines (Fig. 3.1c) (Burt et al. 2012; Krishna et al. 2006; Sam et al. 2006). The virus is transmitted by *Aedes* spp. mosquitoes and exhibits two different transmission cycles. *Aedes aegypti* as well as *Ae. albopictus* are the main vectors in Asia and transmit the virus to humans directly in an urban transmission cycle. Virus isolation from non-human primates or vertebrates like bats in Africa suggests the occurrence of a sylvatic transmission cycle. The main vectors of CHIKV in Africa are *Ae. furcifer-taylori*, *Ae. africanus*, *Ae. luteocephalus* and *Ae. aegypti* (Higgs 2006; Krishna et al. 2006). In the last

50 years, numerous outbreaks of CHIKV were reported (Schwartz and Albert 2010). One of the last large-scale epidemics began in 2004 in Kenya and spread to several islands in the Indian Ocean (Powers and Logue 2007). On La Réunion, nearly 34 % of the total island population was infected and 237 people died (Schwartz and Albert 2010; Tsetsarkin et al. 2007). It was a noteworthy outbreak because *Ae. albopictus* was recognized as the major vector for the first time (Reiter et al. 2006). In July to September 2007, the first autochthonous epidemic outbreak of CHIKV with 248 cases was reported in Italy. The vector responsible for this outbreak was also *Ae. albopictus* (Rezza et al. 2007; Sambri et al. 2008). An infection with CHIKV can cause acute, subacute and chronic diseases. Especially in areas that also suffer outbreaks of Dengue virus, CHIKV can easily be mistaken for Dengue and both viruses can occur in one patient. Dengue has much more potential for causing infections with serious outcomes (PAHO 2011; Tiawsirisup 2011). An acute disease is characterized by fever and joint pain, while other symptoms may include headache, myalgia, nausea, vomiting, polyarthritis, rash and conjunctivitis (PAHO 2011). Apart from humans, bats can also be carriers of the virus. Diallo et al. (1999) isolated it from bats of the genus *Scotophilus* sp. in Senegal. Other surveys suggest *Rousettus aegyptiacus* (Egyptian rousette), *Hipposideros caffer* (Sundevall's leaf-nosed bat) as well as *Chaerephon pumilus* (little free-tailed bat) to be vertebrate hosts for the virus (Calisher et al. 2006). **Ross River virus** (RRV) causes a disease known as epidemic polyarthritis, which is regarded with 2,000–8,000 notified cases per annum as the most common cause of arboviral disease in humans in Australia (Russell 2002; Smith et al. 2011). The first reported outbreak of an infection with RRV was in 1928 during epidemics in New South Wales (Mackenzie et al. 1994; Russell 2002). Today, human infections are also documented for New Guinea, Solomon Islands, Fiji and American Samoa (Meyer 2007; Tesh 1982). It is believed that the virus was introduced to these islands by viraemic air travellers from Australia (Smith et al. 2011). The first isolation of the virus from a mosquito (*Ae. vigilax*) was by Doherty et al. in 1959 (Harley et al. 2001). In 1979, the virus was isolated for the first time from the serum of a patient with epidemic polyarthritis (Mackenzie et al. 1994). The virus was also isolated from a *Pteropus* bat in Australia (Doherty et al. 1966; Messenger et al. 2003). Serological surveys and virus isolation by Harley et al. (2000) from mosquitoes trapped near a flying fox camp suggested the flying fox *Pteropus conspicillatus* as a potential reservoir host. These authors also suggest that other vertebrates living in the flying fox camp could also be important reservoir hosts (Harley et al. 2000, 2001). On the other hand, Ryan et al. (1997) showed that the grey-headed flying fox (*Pteropus poliocephalus*) does not produce a viraemia of sufficient magnitude to be very competent vertebrate hosts of RRV (Ryan et al. 1997). Within 9 years (1991–2000), more than 47,000 laboratory-notified cases were reported by national authorities (Russell 2002). Typical symptoms are various combinations of arthralgia and arthritis, muscle and joint pains, myalgia, lethargy, headache or fever. To return to full physical activity, most of the patients need up to 6 months (Mackenzie et al. 1994; Smith et al. 2011; Weinstein et al. 2011). So far, RRV has been isolated from 27 mosquito species. In Australia,

these include the major vectors *Aedes vigilax*, *Ae. camptorhynchus* and *Cx. annulirostris* (Harley et al. 2000; Hu et al. 2006; Mackenzie et al. 1994). There is also evidence that *Ae. aegypti* can be infected with and transmit the virus, but it has not been isolated from this species in the field (Harley et al. 2000). The **Eastern equine encephalitis virus** (EEEV) is placed as the only species in the Eastern equine encephalitis complex (EEE) and is distributed in North America (Fig. 3.1d) (Calisher et al. 1980). Originally, it was divided into North and South American varieties based on antigenic properties (Casals 1964). Following further antigenic studies four different subtypes have been distinguished, which correspond to four genetic lineages (I–IV) (Arrigo et al. 2010; Brault et al. 1999). EEEV was first recognized as a horse disease in the northeastern USA in 1831 (Hanson 1957; Nalca et al. 2003; Scott and Weaver 1989). The most severe outbreak of EEEV was recorded from Texas and Louisiana in 1947, causing 14,344 cases with 11,722 horse deaths (Chang and Trent 1987; Nalca et al. 2003). However, suspected EEEV could not be linked to humans till an outbreak in 1938 (Fothergill et al. 1938; Getting 1941). In humans, the virus causes severe meningoencephalitis, sometimes causing focal brain lesions. Morse et al. (1992) report the fatality rate in humans of all ages as 30 %, and in affected children up to 75 %. A prodrome of fever, headache, confusion, lethargy, myalgias, vomiting and abdominal pain, which lasts 1–3 weeks, precedes the onset of neurologic symptoms. The onset of illness is characterized by fever, altered mental condition, seizures, vomiting and cyanosis (Deresiewicz et al. 1997; Hart et al. 1964; Romero and Newland 2003). In a study made in Alabama, several mosquito species were found to be carriers of the virus: *Culiseta melanura, Aedes vexans, Coquillettidia perturbans, Culex erraticus* and *Uranotaenia sapphirina*. Interestingly, these species were infected at different times: *Aedes vexans* at the beginning of the season; *Cx. erraticus* and *Cs. melanura* from June till mid-September (Cupp et al. 2003). In most publications, the ornithophilic mosquito *Cs. melanura* is mentioned as the main vector of EEEV, but *Cx. erraticus* is an important bridge vector between birds and mammals in the mid-south USA, because of frequent virus isolations and abundance of this mosquito species in bottomland swamps, flood plains, permanent standing water, recreation areas near rivers or ponds and water impoundments (Jacob et al. 2010). However, EEEV has been isolated from a wide range of species of *Aedes, Anopheles, Coquillettidia, Culex, Ochlerotatus* and *Uranotaenia*, though not all of the species involved can be regarded as competent vectors (more specific e.g. in Armstrong and Andreadis 2010; Arrigo et al. 2010). In surveying for the natural vector of EEEV, Merrill et al. (1934) mention *Ae. cantator* and even more *A. sollicitans* and *A. aegypti*. Regarding the latter species, they conclude with Kelser (1933) that it could not be the transmitting species at it did not occur sufficiently far northwards. Generally, epizootics of EEE occur every 5–10 years and are associated with increased mosquito populations resulting from heavy rainfall and warm weather (Grady et al. 1978; Letson et al. 1993; Mahmood and Crans 1998; Nalca et al. 2003). Although birds seem to be the main reservoir hosts of EEEV, the virus has also been detected in bats. During 1969 and 1970, it was found in ten bat individuals caught in the wild in New Jersey, belonging to three species: one *Lasiurus cinereus*, two *Myotis lucifugus* and seven *Eptesicus fuscus* (Main 1979a). Antibodies have been detected in the previously mentioned species, *Tadarida brasiliensis* and an unidentified bat (Daniels et al. 1960; Hayes et al. 1964; Karstad and Hanson 1958). The survey by Main (1979a) showed that EEE neutralizing antibodies were detectable in a small percentage of the tested adult hibernating animals (0.3 % of *Myotis keenii*) but significantly more in non-hibernating animals (3.4 % in *M. lucifugus, M. keenii* and *E. fuscus*). In experimentally infected bats, the virus could be detected in the blood, mammary glands, brown fat, lung, kidney, brain, pancreas, heart, salivary glands, liver and ovary, with the highest percentage in blood and mammary glands (Main 1979b). In a serological survey of Guatemalan bats, antibodies neutralizing EEEV were found in *Artibeus intermedius, A. jamaicensis, A. lituratus, Glossophaga soricina, Rhynchonycteris naso* and *Sturnira lilium* (Ubico and McLean 1995). The **Venezuelan equine encephalitis virus** (VEEV) belongs to the VEE complex, which consists of six subtypes which have been identified in North, Central and South America (Fig. 3.1e) (Fine et al. 2007; Weaver et al. 1992). The first isolation of VEEV was in 1938 from the brain of a Venezuelan animal (Beck and Wyckhoff 1938). It is the most important pathogen among the New World alphaviruses affecting humans and horses. It not only remains a naturally emerging disease but is also a highly developed biological weapon (Colpitts et al. 2007), because it can be also spread by aerosol infection. Since 1938, sporadic outbreaks have involved hundreds of thousands of people (Weaver et al. 2004). During an epidemic in Colombia, more than 75,000 human cases were reported (Rivas et al. 1997). The symptoms in humans include malaise, sore throats, headaches, fever and chills, persisting for 4–6 days and followed by 2 or 3 weeks of generalized weakness. Encephalitis occurs mainly in children (in about 4 % of cases). Other symptoms range from mild nausea, vomiting with decreased sensorial capability, nuchal rigidity, ataxia and convulsions, to coma and paralysis. The fatality rate in humans is about 1 % (Johnson and Martin 1974; Johnson et al. 1968; Nalca et al. 2003; Pittman et al. 1996). Currently, no vaccine for VEEV is publicly available and the experimental military vaccine has poor efficacy (Colpitts et al. 2007; Russell 1999). In nature, VEEV is maintained in a cycle between mosquitoes and small rodents (Grayson and Galindo 1968; Nalca et al. 2003; Scherer et al. 1972). Epidemic outbreaks of the epizootic strains occur in 10–20-year intervals in the ranch areas in Peru, Venezuela, Colombia and Ecuador after heavy rainfall, which lead to increased mosquito populations (Rivas et al. 1997). Enzootic strains are transmitted by *Culex* species, whereas the main vector of the epizootic strains seems to be *Ochlerotatus taeniorhynchus* and possibly also *Psorophora confinnis*, but nearly all mosquito species have been found to be infected during epizootics (Rivas et al. 1997; Smith et al. 2008; Zacks and Paessler 2010). Weaver et al. (1992) list *Culex cedeci, Cx. ocossa, Cx. panocossa, Cx. portesi, Cx. taeniopus* and cliff swallow bugs *Oeciacus vicarius* as vectors. Other natural enzootic vectors are *Cx. accelerans, Cx. adamesi, Cx. amazonicus, Cx. ferreri, Cx. nigripalpus, Cx. pedroi, Cx. spissipes, Cx. vomerifer, Aedes serratus* and *Mansonia titillans* (Ferro et al. 2003). In the laboratory, *Aedes aegypti* and *Culex aikenii* could additionally be infected (Kramer and Scherer 1976; Sudia et al. 1971). By testing haemagglutination inhibition antibody titres in Guatemalan bats, antibodies against enzootic VEEV strains were detected in seven bat species: *Artibeus jamaicensis*, *A. lituratus*, *A. phaeotis*, *Desmodus rotundus*, *Glossophaga commissarisi*, *Myotis nigricans* and *Uroderma bilobatum*, and it was noted that the insectivorous *M. nigricans* may become infected by eating infected mosquitoes and the vampire bat *D. rotundus* by feeding on viraemic cattle (Seymour et al. 1978a). Experimental infection studies showed that bat genera respond differently to infection, e.g. *Artibeus jamaicensis* and *A. lituratus* showed longer VEEV viraemias than *Phyllostomus discolor*, which had a faster, higher and longer lasting immune response to epizootic strains than to enzootic ones. *Phyllostomus discolor* may not serve as a host for enzootic VEEV strains, while the circulating virus levels in *Artibeus jamaicensis*, *A. lituratus* and *Sturnira lilium* seemed to be high enough to permit the infection of *Culex* vectors (Seymour et al. 1978b).

3.2.3 Flaviviridae: *Flavivirus*

The family Flaviviridae can be divided into the three genera *Pestivirus*, *Hepacivirus* and *Flavivirus* (Cook and Holmes 2006). The last named contains the largest number of viruses with the potential to cause diseases in bats and humans and may be spread by haematophagous insects. *Flavivirus* currently consists of 70 serologically related, single-strand, positive-sense RNA viruses (Hoshino et al. 2009; Mackenzie and Williams 2008; Tajima et al. 2005). All members of the genus can be found around the world and are most often transmitted by arthropods. The most important mosquito-borne diseases caused by flaviviruses are Yellow fever, Japanese Encephalitis, West Nile fever, Dengue fever and St. Louis encephalitis (Mackenzie and Williams 2008). **Yellow fever** is an acute haemorrhagic disease and is endemic in tropical areas of Africa and Latin America with an incidence of 200,000 cases and 30,000 deaths each year (Fig. 3.2b). The symptoms range from mild to severe illness (WHO 2000). Yellow fever transmission depends on arthropods as vectors, e.g. several mosquito (Culicidae) species belonging to *Aedes* in Africa and *Haemogogus* in South America (Ellis and Barrett 2008; WHO 2000). As long ago as the first half of the twentieth century, scientists tried to detect a connection between the yellow fever virus and bats. Williams et al. (1964) emphasized the potential and importance of bats in the wild cycle of arboviruses, e.g. yellow fever virus. In the experiments of Kumm (1932), Brazilian bats (*Molossus molossus obscurus* and *M. rufus*) were exposed to unfed *Aedes aegypti*. The mosquitoes willingly took blood, but no transmission of the disease from the infected mosquitoes to the bats resulted. The author concluded that bats play little part in the life cycle of this disease. Contrastingly, Simpson and O'Sullivan (1968) stated that in East Africa, yellow fever virus circulated in the fruit bat genera *Eidolon* and *Rousettus*. However, the virus did not produce demonstrable viraemia in the tested fruit bats (Simpson and O'Sullivan 1968). Marinkelle

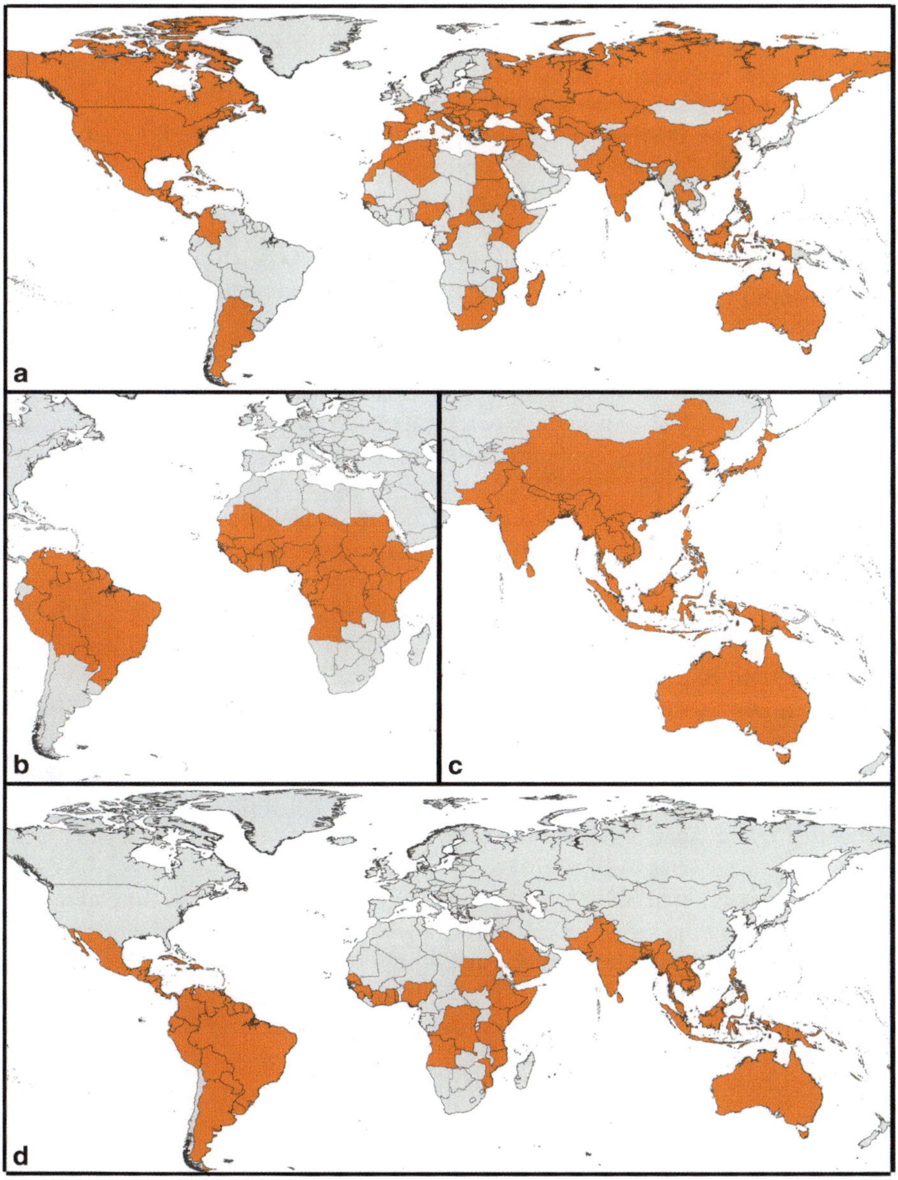

Fig. 3.2 Occurrence of viruses (**a**) West Nile virus: detected in human sera or antibodies in birds (Hubálek and Halouzka 1999; Gubler 2007); (**b**) Yellow fever: *dark orange* (do)—risk area (Travel approved 2010); (**c**) Japanese encephalitis virus (CDC 2012; Van der Hurk et al. 2009b); (**d**) Dengue fever: transmission risk areas (WHO 2009)

and Grose (1972) reviewed organisms, which act as vectors between bats, humans and domestic animals. They listed six references involving yellow fever records from two continents.

Oelofsen and van der Ryst (1999) stated that bats can be infected orally by ingesting a single mosquito and that experiments with bats and yellow fever virus produced positive results. The virus was recovered from several organs of the bats (Oelofsen and van der Ryst 1999). Furthermore, serological prevalence of bats for yellow fever virus had been demonstrated in Uganda, Kenya and Sudan and in previous publications the laboratory capacity had been tested in the genera *Eidolon* and *Rousettus* (Ellis and Barrett 2008). The **Japanese encephalitis virus** (JEV) is eponymous for the JEV antigenic complex. Other members of this group are the West Nile virus, Murray Valley encephalitis virus, St. Louis encephalitis virus and Kunjin virus (Bengis et al. 2004; Campbell et al. 2002). It is the leading cause of viral encephalitis in rural regions of eastern, southeastern and southern Asia (Fig. 3.2c). The central nervous system may be affected, leading to severe complications and even death (Agarwal 2006; Mackenzie et al. 2006). Up to 50,000 cases of JEV are estimated to occur annually worldwide (Ravanini et al. 2012; Van der Hurk et al. 2009a). According to the sequence of its genomic RNA, JEV is classified into five genotypes (Nabeshima et al. 2009; Solomon et al. 2003). The majority of infections are subclinical, but the fatality rate is nearly 25 % (Ravanini et al. 2012; Van der Hurk et al. 2009b). Over the last 60 years, it has been estimated that the virus has been responsible in humans for more than ten million infections, three million deaths and four million cases of long-term disability. It is also calculated that today nearly two billion people live in JEV-prone areas. Historically, epidemics had been recorded in Japan since 1871 (Mackenzie et al. 2006; Van der Hurk et al. 2009a), but the first isolation of JEV in Japan was not until 1935 in Japan (Tiawsisirsup et al. 2012). A sequence closely related to JEV strains from Japan was isolated for the first time in Europe from the mosquito *Culex pipiens* (Ravanini et al. 2012). Birds and mosquitoes play the major role in the life cycle of JEV, but the virus can also infect a wide range of other vertebrates such as humans, domestic animals, bats, snakes or frogs (Agarwal 2006; Ravanini et al. 2012; Tiawsisirsup et al. 2012). The main vectors seem to be species of *Culex* mosquitoes such as *Cx. pipiens* (Korea), *Cx. annulirostris* (Australia) or *Cx. tritaeniorhynchus* (Nabeshima et al. 2009; Van der Hurk et al. 2009b), but JEV has also isolated from *Cx. gelidus*, *Cx. vishnui*, *Cx. fucocephala*, *Cx. pseudovishnui*, *Mansonia uniformis*, *Anopheles subpictus* and *Ochlerotatus japonicus* (Mackenzie et al. 2006; Van der Hurk et al. 2009a). Sulkin et al. (1970) isolated the virus in Japan from the bats *Miniopterus schreibersii* and *Rhinolophus cornutus*. Other bats, which have yielded antibodies or JEV, were *Myotis mystacinus*, *Pipistrellus abramus*, *Plecotus auritus*, *Vespertilio superans*, *Myotis macrodactylus* and *M. nattereri bombinus* (Banerjee et al. 1984). In China, JEV has been isolated from *Rousettus lechenaultii* and *Murina aurata* (Wang et al. 2009). **West Nile virus**, the disease causing agent of the West Nile fever, was first discovered in the blood of a native woman of the West Nile district of Uganda in 1937 (Garmendia et al. 2001; George et al. 1984). In South Africa, one of the biggest outbreaks, with nearly 18,000 human infections, was reported in 1974 (Dauphin et al. 2004). Molecular epidemiological survey indicates that WNV spread from Africa to the Mediterranean and southern European regions and then to India as well as Central and South Asia (Fig. 3.2a) (Buckley et al. 2003; Hayes 2006). In 1999, the virus was inadvertently introduced into North America (Pilipski et al. 2004). Up to now, over 12,000 human cases of meningitis or encephalitis and 1,100 deaths caused by WNV have been documented in the USA (Murray et al. 2010). Phylogenetic studies have identified several genetic lineages of the virus in different geographical locations (Campbell et al. 2002; Rappole et al. 2000). West Nile virus is transmitted in natural cycles between birds and mosquitoes (Mackenzie et al. 2004). So far, the virus has been isolated from at least 300 bird and 43 mosquito species from 11 genera. The major vectors for Africa and the Middle East are *Culex univittatus, Cx. poicilipes, Cx. neavei* or *Aedes albocephalus*. For Asia, it is *Cx. quinquefasciatus, Cx. tritaeniorhynchus* and *Cx. vishnui*. In Europe, *Cx. pipiens, Cx. modestus* as well as *Coquillettidia richardii* may act as important vectors. A huge number of infections (nearly 80 % symptomless) occur during summer, early fall and during the rainy season in the tropics (Campbell et al. 2002). The symptoms reach from fever, headache, tiredness or swollen lymph glands to neck stiffness, disorientation, coma and paralysis (Mackenzie et al. 2004; WHO 2011; Zeller and Schuffenecker 2004). The first identification of WNV in Chiroptera was in *Rousettus aegyptiacus* (fruit bat) in Uganda and Israel. Nearly 8 % of the surveyed *R. aegyptiacus* in Israel tested positive for WNV antibodies (Bunde et al. 2006). In India, the virus was isolated from *Rousettus leschenaultia* (Davis et al. 2005; Paul et al. 1970) and in 2,000 antibodies were isolated again from live *Eptesicus fuscus* (big brown bat) as well as from *Myotis lucifugus* (little brown bat) in New York City (Bunde et al. 2006). Pilipski et al. (2004) found antibodies again in *M. lucifugus* and in *M. septentrionalis*, whereas Davis et al. (2005) determined neutralizing antibodies for WNV from *Tadarida brasiliensis*. The **St. Louis encephalitis** virus is the etiological agent of St. Louis encephalitis and is a member of the Japanese encephalitis antigenic complex. It was first detected in 1933 during an outbreak of human encephalitis in St. Louis in the US State Missouri (Auguste et al. 2009; Flores et al. 2010; Rodrigues et al. 2010). Today, the virus is found all over the USA and Canada as well as Central and South America (Diaz et al. 2006; Pires and Gleiser 2010). The largest outbreak of SLEV among humans so far was in 2005 in Argentina (Diaz et al. 2006). The first detection of SLEV in Argentina was in 1957 (Flores et al. 2010). Studies have indicated *Cx. tarsalis*, *Cx. nigripalpus* and *Cx. quinquefasciatus* to be the major vectors in the USA (Reisen 2003). Diaz et al. (2012) isolated the virus from eight different mosquito species: *Cx. quinquefasciatus*, *Cx. interfor*, *Cx. apicinus*, *Ae. scapularis*, *Ae. aegypti*, *Ae. albifasciatus*, *An. albitarsis* and *Ps. ferox*. The primary transmission cycle is between mosquitoes and birds, but serological evidence of infection has also been found in horses, cattle and goats (Calisher 1994; Spinsanti et al. 2003). In a survey by Bunde et al. (2006), the bats *Eptesicus fuscus* (big brown bat) and *Myotis lucifugus* (little brown bat) were tested positive for SLEV antibodies. An infection with SLEV can cause a slight illness with fever and headache or serious illness with meningoencephalitis and death. **Dengue** is a viral infection and characterized by symptoms like fever, severe headache, orbital pain and general indisposition and start 5–7 days after infection. Haemorrhages and an increase in vascular permeability are the consequences of the Dengue haemorrhagic fever, which frequently leads to death. There is good evidence that sequential infection with different serotypes increases the risk of developing this dangerous form of dengue fever (Aguilar-Setién et al. 2008; Becker et al. 2010; Kalayanarooj et al. 1997). Dengue infections increased worldwide during the last decades and about one-fifth of the world population lives in Dengue risk zones (Fig. 3.2d) (Thomas et al. 2011). Dengue virus has been detected worldwide in tropical and subtropical regions, especially not only in the Southeast and South Asia but also in Central and South America and with an ongoing transmission risk in Africa (e.g. Chen and Wilson 2005). The first reported cases occurred at the end of the eighteenth century in Asia, Africa and North America, while Dengue haemorrhagic fever first occurred in the 1950s in the Philippines and Thailand (Becker et al. 2010). In the first half of the twentieth century in Europe (Austria, Greece, Italy and Spain), the virus caused epidemics. Primary vectors are species of the mosquito genus *Aedes* (*Aedes aegypti* and *Ae. albopictus*). Hypothetically, insectivore bats might become infected by ingestion of virus-infected mosquitoes, while fructivore species have to be infected by a mosquito bite (Aguilar-Setién et al. 2008). Wong et al. (2007) categorized the risk of bat to human transmission for the families Pteropodidae and Phyllostomidae as low because of low prevalence of pathogens in bats or inefficient vectorial capacity. However, de Thoisy et al. (2009) detected dengue viral RNA in 4 % of Chiroptera samples (*Carollia perspicillata*) from French Guiana. Platt et al. (2000) detected antibodies against Dengue virus in 22.6 % of examined bats from Costa Rica ($n = 53$) and 30 % of those from Ecuador ($n = 10$), mainly not only in bats of the genera *Artibeus* and *Uroderma* but also in four species of *Molossus*. In laboratory experiments, *Ae. aegypti* from Costa Rica fed on bats (Platt et al. 2000), but Scott (2001) doubted that this feeding is consistent with bat involvement in Dengue transmission and considered that there is no proof of interactions in natural conditions. In Mexico, Dengue virus is transmitted between humans by mosquitoes of the genus *Aedes*. Bat samples ($n = 162$) from five families (Emballonuridae, Mormoopidae, Phyllostomidae, Natalidae and Vespertilionidae) contained nine individuals of four species that were seropositive according to ELISA (*Artibeus jamaicensis, Myotis nigricans, Pteronotus parnellii* and *Natalus stramineus*). This is the first definite evidence of Dengue virus in *Myotis* species. These results support the contention that Dengue virus is present in bats from the Pacific and gulf coasts of Mexico (Aguilar-Setién et al. 2008). The first isolation of **Ilheus virus** (ILHV) was from mosquitoes of the genera *Ochlerotatus* and *Psorophora* from Brazil, especially *Psorophora ferox*, which is considered its main vector (da Silva Azevedo et al. 2010; Laemmert and Hughes 1947). Later it was isolated also from the genera *Culex, Haemagogus, Sabethes* and *Trichoprospon* (Venegas et al. 2012). In Brazil, it has been isolated also from birds, sentinel monkeys and horses (Iversson et al. 1993). Only a few reports of isolation from humans are available (Johnson et al. 2007; Spence et al. 1962; Srihongse and Johnson 1967;

Venegas et al. 2012). Results of infection are widely variable, ranging from asymptomatic to encephalitis, but most of the cases are accompanied by fever, headache, chills, photophobia, arthralgia, myalgia and asthenia (da Silva Azevedo et al. 2010). Price (1978) found sera in bats from Trinidad that protect against Ilheus. According to da Silva Azevedo et al. (2010), it has been isolated also from bats, but unfortunately no details of which species were involved were mentioned by the authors. **Zika virus** is known from Africa and Southeast Asia (Dick et al. 1952; Duffy et al. 2009; Hayes 2009). It is related to West Nile, Dengue and Yellow fever viruses (Duffy et al 2009). The first isolation of Zika virus was in 1947 from a rhesus monkey (Simpson 1964) in 1948 was the first isolation from a mosquito (*Aedes africanus*) and in 1968 from humans in Nigeria (Hayes 2009). Other serological studies showed human Zika virus infection in Africa also e.g. in Senegal, Uganda, Central African Republic and Egypt and for Asia e.g. in India, Malaysia, Vietnam or Indonesia (Duffy et al. 2009; Hayes 2009). Zika virus has been isolated from *Ae. aegypti, Ae. africanus, Ae. furcifer, Ae. luteocephalus* and *Ae. vittifer* (Dick 1952; Haddow et al. 1964; Hayes 2009; Lee and Moore 1972; Marchette et al. 1969). So far, no natural infections of bats with Zika virus have been documented, but in the laboratory the cave bat *Myotis lucifugus* was infected successfully when the virus was injected intraperitoneal, intradermal, intracerebral and intrarectal. But the bats were not susceptible to the virus after intranasal exposure (Reagan et al. 1955). **Tacaribe virus** (TCRV) belongs to the Arenaviridae (genus *Arenavirus*) (Bowen et al. 1996; Rossi et al. 1996). Diseases caused by the Tacaribe virus complex of the new world are Argentine haemorrhagic fever, Brazilian haemorrhagic fever, Venezuelan haemorrhagic fever and another yet unnamed haemorrhagic fever, induced by Junín, Sabiá, Guanarito, Machupo and Chapare viruses (Carballal et al. 1987; Cogswell-Hawkinson et al. 2012; Tesh et al. 1994). In contrast to the other arenaviruses, which have all been isolated from rodents, Tacaribe virus was originally isolated from two bat species: great fruit-eating bats (*Artibeus lituratus*) and Jamaican fruit bats (*A. jamaicensis*) (Downs et al. 1963; Price 1978). Furthermore, Price (1978) was able to detect antibodies against TCRV in the little yellow-shouldered bat (*Sturnira lilium*), Heller's broad-nosed bat (*Platyrrhinus helleri*) and in the vampire bat (*Desmodus rotundus*). Nevertheless, the study of Cogswell-Hawkinson et al. (2012) does not support the hypothesis that *A. jamaicensis* is a natural reservoir host for TCRV, because the injection of high doses resulted in significant and fatal disease including pneumonia, pathological changes in liver and spleen and brain lesions. One study (Downs et al. 1963) however revealed close correspondence between a strain isolated from a mosquito pool and one isolated from a bat. Unfortunately, the pool consisted of 18 mosquito species, so that no precise information about the possible vector can be given.

Besides viruses, some parasites are known, which occur in bats and humans, and can be transmitted through haemorrhagic insects. The first one is the **Chagas disease** an infection with the flagellate *Trypanosoma cruzi* (Zeledón and Rabinovich 1981). The disease occurs in Central and South America and is transmitted by the intestinal content of triatomine kissing bugs (Mehlhorn 2001). It is the leading cause of heart disease in South America with the major vector species *Triatoma infestans*, *T. dimidiata* and *Rhodnius prolixus* (Reduviidae) (Dorn et al. 2003). Flagellates of the genus *Trypanosoma* are parasitic in nearly all mammalian species. In more than 100 species of bats more than 30 trypanosome species are recorded, while the subgenus *Schizotrypanum* comprises species restricted to bats as well as *T. cruzi* (Cavazzana et al. 2010). Kissing bugs live for example in caves, burrows, nests of wild animals on which they feed during the night. During the blood meal infected faeces is set free and after the bite scratched by the victim in the itching wound. During the life cycle in man or other reservoir hosts, amastigotes reproduce in the cytoplasm of different host cell types, which appear as "pseudocysts" when they are completely filled with parasites. The most important lesions are in the heart and a myocardial failure results to death years after the infection (Mehlhorn 2001). The symptoms range from fever to inflammation of heart, muscles and brain (CDC 2010).

Beneath the usual triatomine vector, several wild animals are associated with the Chagas disease. Among others Marsupialia, Rodentia, Edentata, Primates and Chiroptera had been positive investigated for *T. cruzi* (Coura et al. 2002). Common species of neotropical bats, including those of the genera *Artibeus*, *Noctilio*, *Mormoops*, *Nautilus*, *Pteronotus*, *Myotis*, *Carollia*, *Desmodus*, *Glossophaga*, *Phyllostomus* and *Molossus*, have been reported to be susceptible to *T. cruzi* infection under natural as well as under experimental conditions (Añez et al. 2009). Bats can get infected by the blood meal of the kissing bugs or through the ingestion of infected arthropods. So it is not astonishing that most infected bats are insectivorous. The prevalence in South American bats varied widely. In Colombia and in the Amazonia of Brazil, it is approximately 9.0 %, respectively 2.4–4.6 % (García et al. 2012).

Añez et al. (2009) detected in *Molossus molossus* in western Venezuela a congenital transmission from pregnant female bats to their foetus. Trypomastigotes had been found in 100 % of all examined foetus. In their natural habitat, *M. molossus* is associated with *R. prolixus* kissing bugs. The insectivore *M. molossus* shows a high susceptibility for *T. cruzi*, due to the fact that 80 % of the examined bats are infected. These results emphasize the role of Chiroptera as host for Chagas disease in endemic areas and their impact for the sylvatic cycle of *T. cruzi* (Añez et al. 2009).

Recent examinations detected new genotypes of *T. cruzi* associated with bats, which indicate that the complexity of *T. cruzi* is larger than known and confirmed bats as important reservoir for infections to humans (e.g. Maeda et al. 2011; Marcili et al. 2009) and the strong association between bats and, for instance, *Schizotryphanum* suggests a long shared evolutionary development (García et al. 2012). Furthermore, the molecular examination of Chagas virus strains reveals the movement of bats, naturally or by human transport, between the Old and the New World (Hamilton et al. 2012b). Hamilton et al. (2012a) suggested that *T. cruzi* evolved from bat trypanosomes and have successful switched into other mammalian hosts.

Coccidia (Apicomplexa) are characterized by intracellular life cycles consisting of the three phases: schizogony, gamogony and sporogony. The coccidian genus ***Plasmodium*** is the pathogenic agent of malaria, a mosquito-borne infectious disease of humans and animals, which causes fever, headache and in severe cases death (Mehlhorn 2001). Worldwide 3.3 billion people live in risk areas (Africa, Southeast Asia region and the Eastern Mediterranean) of malaria transmission and each year at least one million people die after infection (Snow et al. 2005). The order Haemosporidia consists of the five genera: *Plasmodium, Hepatocystis, Polychromophilus, Nycteria and Rayella.* The vectors of the first three genera are respectively haematophagous Diptera of the families Culicidae (*Anopheles* spp.), Ceratopogonidae and Nycteribiidae (Witsenburg et al. 2012). With the exception of *Rayella*, all haemosporidia genera are known to infect insectivorous bats in temperate and tropical regions (Duval et al. 2012). Megali et al. (2011) investigated 237 bats of four species from Switzerland to obtain a better understanding of the complex co-evolutionary processes between hosts and parasites. A total of 34 % ($n = 70$) was infected with *Plasmodium murinus*. In detail, *Myotis daubentonii* was the most parasitized species (51 %), followed by *Eptesicus serotinus* (11 %), *Nyctalus noctula* (7 %) and *Myotis myotis* (4 %) (Megali et al. 2011). The prevalence (P) of *P. murinus* in *M. daubentonii* was twice as great as found by Gardner and Molyneux (1988) in England and Scotland. Duval et al. (2007) examined 530 bat individuals (Pteropodidae, Rhinolophidae, Hipposideridae, Megadermatidae, Emballonuridae, Vespertilionidae and Molossidae) from Madagascar and Cambodia. In Madagascar haemosporidian infections were found in Hipposideridae (*Triaenops furculus*, P 4 %) and Vespertilionidae (*Miniopterus gleni*, P 23 %; *Myotis goudoti*, P 24 % and *Miniopterus manavi*, P 38 %). In Cambodia, infections were found in the Hipposideridae (*Hipposideros larvatus*, P 8 %), Megadermatidae (*Megaderma spasma*, P 80 %) and Vespertilionidae (*Kerivoula hardwickii*, P 20 %) (Duval et al. 2007). In *Pteropus poliocephalus* from Australia (P 36 %), Landau et al. (1980) described *Hepatocystis levinei*, which under laboratory conditions, used *Culicoides nubecolusus* (Diptera and Ceratopogonidae) to complete its life cycle. Landau et al. (2012) listed bat hosts of six families in which different types of gametocytes were detected. The authors concluded that the Microchiroptera harbour mainly parasites of the *falciparum* and *malariae* groups, while Megachiroptera harbour parasites of the *vivax* group (Landau et al. 2012). Duval et al. (2012) sampled 164 bats from Gabon in Central Africa of which only *Miniopterus inflatus* was positively tested for haemosporidian parasites. The prevalences ranged from 17.6 % to 66.7 % (blood smear examinations), while the molecular prevalence ranged from 63.2 % to 88.9 %. The nycteribiid *Polychromophilus fulvida* was found infected with *Polychromatophilia* sp. in Faucon Cave in Gabon (Landau et al. 1980). *Miniopterus inflatus* and other bat species (e.g. *H. gigas, H. caffer* and *C. afra*) are potentially exposed to this blood parasite (Duval et al. 2012). The five host specific *Polychromophilus* species are restricted, regarding their vertebrate hosts, to insectivorous bats of the order Microchiroptera. Vectors are Nycteribiidae (Diptera, Hippoboscoidea) (Witsenburg et al. 2012).

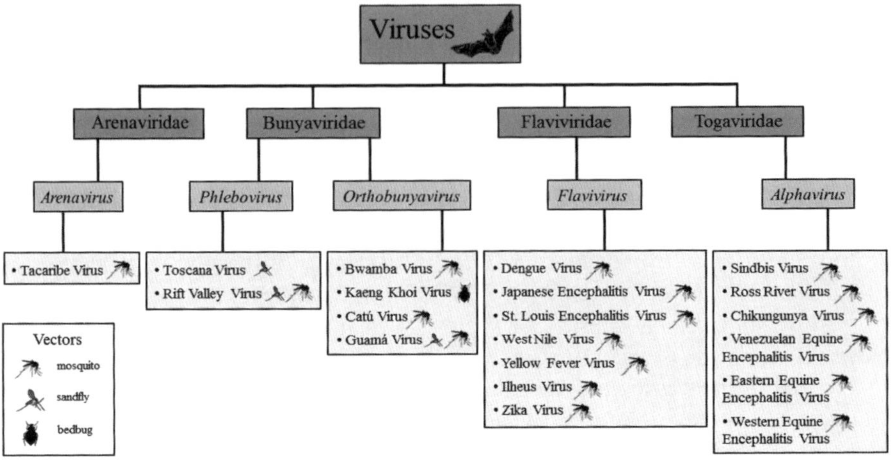

Fig. 3.3 Overview of the pathogens found in bats and human with their family, genera and vectors, which are discussed in this chapter

3.3 Conclusion

In the past, a high number of viruses and parasites has been detected in bats, which are important reservoir hosts. On the contrary, many haematophagous insects serve as vectors for numerous arboviruses and parasites, with mosquitoes (Culicidae) being the most important vectors worldwide. So the presence of certain viruses in both mosquitoes and bats is not really surprising and a transmission cycle between bats, mosquitoes and humans is thinkable. However, it is not possible to say whether the mosquitoes served as vector for the bats or the bats as reservoir both options must be taken into consideration.

In the present chapter, we describe 20 viruses from four different families as well as two parasitic pathogens, which have been detected in bats, in haematophagous insects and in humans (see also Table 3.1, Fig. 3.3). Therefore, for these pathogens is a probability to be transmitted from bats to humans via insects. However, because of the amount of publications in these fields and the quantity of described viruses it is difficult to give a definite number for the possible diseases and the chapter makes no claim to be complete. So far, this way of transmission between bats–mosquito–humans could not be proven for any of the diseases, but vectors being in relation with bat infections (e.g. *Cx. quinquefasciatus*, *Ae. vexans* or *Ae. aegypti* and *Ae. albopictus*) are also known to bite humans. Climate change as well as global trade could not only increase the risk for such way of transmission e.g. when potential hosts but also competent vectors expand their distribution. Apart from the listed viruses and parasites (see Table 3.1, Fig. 3.3) are some pathogens often described only in humans or bats but has been detected already also in mosquitoes or is, at least, suspected to be transmitted by arthropods, like Bimiti, Oriboca, Mayaro or Yokose virus. With climate change range swift of vectors and possible spontaneous mutations, new hosts and/or vector competences can occur. For some diseases like WNV, Dengue or Yellow fever, the way of transmission is already well known, while this is not the case for others like Kaeng Khoi, Catu, Guama, Zika or Bwamba virus, but knowledge about these things is essential as epidemics cause high social as well as economic impact.

Acknowledgements The present study was financially supported by the research funding programme "LOEWE—Landes-Offensive zur Entwicklung Wissenschaftlich-ökonomischer Exzellenz" of Hesse's Ministry of Higher Education, Research, and the Arts and by the SAW (Senate Competition Committee) grant (SAW-2011-BNI-3) of the Leibniz Association.

References

Agarwal AK (2006) Post viral encephalitis sequelae and their rehabilitation. Indian J Phys Med Rehabil 17:39–40

Aguilar-Setién A, Romero-Almaraz ML, Sanchez-Hernández C, Figuero R, Juárez-Palma LP, Garcia-Flores MM, Vázquez-Salinas C, Salas-Rojas M, Hidalgo-Martinez AC, Aguilar Pierle S, Garcia-Estrada C, Ramos C (2008) Dengue virus in Mexico bats. Epidemiol Infect 136: 1678–1683

Añez N, Crisantea G, Soriano PJ (2009) *Trypanosoma cruzi* congenital transmission in wild bats. Acta Trop 109:78–80

Armstrong PM, Andreadis TG (2010) Eastern equine encephalitis virus in mosquitoes and their role as bridge vectors. Emerg Infect Dis 16:1869–1874

Arrigo NC, Adams AP, Weaver SC (2010) Evolutionary patterns of eastern equine encephalitis virus in North versus South America suggest ecological differences and taxonomic revision. J Virol 84:1014–1025

Auguste AJ, Pybus OG, Carrington CVF (2009) Evolution and dispersal of St. Louis encephalitis virus in the Americas. Infect Genet Evol 9:709–715

Ayers JR, Lester TL, Angulo AB (1994) An epizootic attributable to western equine encephalitis virus infection in emus in Texas. J Am Vet Med Assoc 205:600–601

Baldelli F, Ciufolini MG, Francisci D, Marchi A, Venturi G, Fiorentini C, Luchetta ML, Bruto L, Pauluzzi S (2004) Unusual presentation of life-threatening Toscana virus Meningoencephalitis. Clin Infect Dis 38:515–520

Balvín O (2008) Revision of the West Palaerarctic *Cimex* species. Preliminary report. Bull Insectol 61:129–130

Balvín O, Munclinger P, Kratochvil L, Vilímová J (2012) Mitochondrial DNA and morphology show independent evolutionary histories of bedbug *Cimex lectularius* (Heteroptera: Cimicidae) on bats and humans. Parasitol Res 111:457–469

Banerjee K, Ilkal MA, Deshmukh PK (1984) Susceptibility of *Cynopterus sphinx* (frugivorus bat) & *Suncus murinus* (house shrew) to Japanese encephalitis virus. Indian J Med Res 79:8–12

Beck CE, Wyckhoff RWG (1938) Venezuelan equine encephalomyelitis. Science 88:530

Becker N, Petrić D, Zgomba M, Boase C, Dahl C, Lane J, Kaiser A (2010) Mosquitoes and their control. Kluwer Academic/Plenum, New York, NY, 578 pp

Bengis RG, Leighton FA, Fischer JR, Artois M, Mörner T, Tate CM (2004) The role of wildlife in emerging and re-emerging zoonoses. Rev Sci Tech 23:497–511

Blackburn NK, Foggin CM, Searle L, Smith PN (1982) Isolation of Sindbis virus from bat organs. Cent Afr J Med 28:201

Bouloy M (2011) Molecular biology of phleboviruses. In: Plyusnin A, Elliott RM (eds) Bunyaviridae, molecular and cell biology. Caister Academic, Norwich, pp 95–128

Bowen MD, Peters CJ, Nichol ST (1996) The phylogeny of New World (Tacaribe complex) arenaviruses. Virology 219:285–290
Bowen MD, Trappier SG, Sanchez AJ, Meyer RF, Goldsmith CS, Zaki SR, Dunster LM, Peters CJ, Ksiazek TG, Nichol ST, Task Force RVF (2001) A reassortant Bunyavirus isolated from acute hemorrhagic fever cases in Kenya and Somalia. Virology 291:185–190
Braito A, Corbisiero R, Corradini S, Marchi B, Sancascian N, Fiorentini C, Ciufolini MG (1997) Evidence of Toscana virus infections without central nervous system involvement: a serological study. Eur J Epidemiol 13:761–764
Brault AC, Powers AM, Chavez CL, Lopez RN, Cachon MF, Gutierrez LF, Kang W, Tesh RB, Weaver SC, Carrington CVF (1999) Genetic and antigenic diversity among eastern equine encephalitis viruses from North, Central and South America. Am J Trop Med Hyg 61:579–586
Buckley A, Dawson A, Moss SR, Hinsley SA, Bellamy PE, Gould EA (2003) Serological evidence of West Nile virus, Usutu virus and Sindbis virus infection of birds in the UK. J Gen Virol 84:2807–2817
Bunde JM, Heske EJ, Mateus-Pinilla NE, Hofmann JE, Novak RJ (2006) A survey for West Nile virus in bats from Illinois. J Wildl Dis 42:455–458
Burt FJ, Rolph MS, Rulli NE, Mahalingam S, Heise MT (2012) Chikungunya: a re-emerging virus. Lancet 379:662–671
Burton GJ (1963) Bedbugs in relation to transmission of human diseases. Review of the literature. Public Health Rep 78:513–524
Calisher CH (1994) Medically important arboviruses of the United States and Canada. Clin Microbiol Rev 7:89–116
Calisher CH, Shope RE, Brandt W, Casals J, Karabatsos N, Murphy FA, Tesh RB, Wiebe ME (1980) Proposed antigenic classification of registered arboviruses I. Togaviridae, *Alphavirus*. Intervirology 14:229–232
Calisher CH, Childs JE, Field HE, Holmes KV, Schountz T (2006) Bats: important reservoir hosts of emerging viruses. Clin Microbiol Rev 19:531–545
Campbell GL, Marfin AA, Lanciotti RS, Gubler DJ (2002) West Nile virus. Lancet Infect Dis 2:519–529
Carballal G, Calello MA, Laguens RP, Weissenbacher MC (1987) Tacaribe virus: a new alternative for Argentine hemorrhagic fever vaccine. J Med Virol 23:257–263
Casals J (1964) Antigenic variants of eastern equine encephalitis virus. J Exp Med 119:547–565
Causey OR, Causey CE, Maroja OM, Macedo DG (1961) The isolation of arthropod-borne viruses, including members of two hitherto undescribed serological groups, in the Amazon region of Brazil. Am J Trop Med Hyg 10:227–249
Cavazzana Jr. M, Marcili A, Lima L, Maia da Silva F, Junqueira ACV, Veludo HH, Viola LB, Campaner M, Nunes VLB, Paiva F, Coura JR, Camargo EP, Teixeira MMG (2010) Phylogeographical, ecological and biological patterns shown by nuclear (ssrRNA and gGAPDH) and mitochondrial (Cyt b) genes of trypanosomes of the subgenus *Schizotrypanum* parasitic in Brazilian bats. Int J Parasitol 40:345–355
CDC (2002) Summary of notifiable disease-United States, 2000. MMWR Morb Mortal Wkly Rep 49:1–100
CDC (2010) Parasites-American trypanosomiasis (also known as Chagas disease). http://www.cdc.gov/parasites/chagas/disease.html
CDC (2012) Geographic distribution of Japanese encephalitis virus. http://www.cdc.gov/japaneseencephalitis/maps/index.html
Chakravarty AK, Sarkar SK (1994) Immunofluorescence analysis of immunoglobulin bearing lymphocytes in the Indian fruit bat: *Pteropus giganteus*. Lymphology 27:97–104
Chang G-JJ, Trent DW (1987) Nucleotide sequence of the genome region encoding the 26S mRNA of eastern equine encephalomyelitis virus and the deduced amino acid sequence of the viral structural proteins. J Gen Virol 68:2129–2142
Chapman AD (2009) Numbers of living species in Australia and the world. Australian Biological Resources Study, Canberra, 60 pp

Charrel RN, Gallian P, Navarro-Marí J-M, Nicoletti L, Papa A, Sánchez-Seco MP, Tenorio A, de Lamballerie X (2005) Emergence of Toscana virus in Europe. Emerg Infect Dis 11:1657–1663
Chen LH, Wilson ME (2005) Non-vector transmission of Dengue and other mosquito-borne flaviviruses. Dengue Bull 29:18–31
Cogswell-Hawkinson A, Bowen R, James S, Gardiner D, Calisher CH, Adams R, Schountza T (2012) Tacaribe virus causes fatal infection of an ostensible reservoir host, the Jamaican fruit bat. J Virol 86:5791–5799
Colpitts TM, Moore AC, Kolokoltsov AA, Davey RA (2007) Venezuelan equine encephalitis virus infection of mosquito cells requires acidification as well as mosquito homologs of the endocytic proteins Rab5 and Rab7. Virology 369:78–91
Constantine DG (1970) Bats in relation to the health, welfare and economy of man. In: Winsatt WE (ed) Biology of bats. Academic, New York, pp 319–449
Cook S, Holmes EC (2006) A multigene analysis of the phylogenetic relationships among the flaviviruses (Family: Flaviviridae) and the evolution of vector transmission. Arch Virol 151:309–325
Coura JR, Junqueira ACV, Fernandes O, Valente SAS, Miles MA (2002) Emerging Chagas disease in Amazonian Brazil. Trends Parasitol 18:171–176
Cupp EW, Klingler K, Hassan HK, Viguers LM, Unnasch TR (2003) Transmission of eastern equine encephalomyelitis virus in central Alabama. Am J Trop Med Hyg 68:495–500
Cusi MG, Savellini GG, Zanelli G (2010) Toscana virus epidemiology: from Italy to beyond. Open Virol J 4:22–28
da Silva Azevedo RS, Barros VLRS, Martins LC, Ribeiro Cruz AC, Rodrigues SG, da Costa Vasconcelos PF (2010) Pathogenesis of the Ilheus virus in golden hamsters (*Mesocricetus auratus*). Rev Pan-Amaz Saude 1:73–80
Daniels JB, Stuart G, Wheeler RE, Gifford C, Ahearn JP, Philbrook R, Hayes RO, MacCready RA (1960) A search for encephalitis and rabies in bats of eastern Massachusetts. N Engl J Med 263:516–520
Darai G, Handermann M, Sonntag HG, Zöller L (2011) Lexikon der Infektionskrankheiten des Menschen: Erreger, Symptome, Diagnose. Springer, Berlin, 130 pp
Daubney R, Hudson JR (1931) Enzootic hepatitis or Rift Valley fever. An undescribed virus disease of sheep, cattle and man from East Africa. J Pathol 34:545–579
Dauphin G, Zientara S, Zeller H, Murgue B (2004) West Nile: worldwide current situation in animals and humans. Comp Immunol Microbiol Infect Dis 27:343–355
Davis A, Bunning M, Gordy P, Panella N, Blitvivh B, Bowen R (2005) Experimental and natural infection of North American bats with West Nile virus. Am J Trop Med Hyg 73:467–469
de Thoisy B, Lacoste V, Germain A, Muñoz-Jordán J, Colón C, Mauffrey J-F, Delaval M, Catzeflis F, Kazanji M, Matheus S, Dussart P, Morvan J, Setién AA, Deparis X, Lavergne A (2009) Dengue infection in neotropical mammals. Vector Borne Zoonotic Dis 9:157–169
Delaunay P, Blanc V, Del Giudice P, Levy-Bencheton A, Chosidow O, Marty P, Brouqui P (2011) Bedbugs and infectious diseases. Clin Infect Dis 52:200–210
Depaquit J, Grandadam M, Fouque F, Andry PE, Peyrefitte C (2010) Arthropod-borne viruses transmitted by phlebotomine sandflies in Europe: a review. Euro Surveill 15:19507
Deresiewicz RL, Thaler SJ, Hsu L, Zamani AA (1997) Clinical and neuroradiographic manifestations of eastern equine encephalitis. N Engl J Med 336:1867–1874
Diallo M, Thonnon J, Traore-Lamizana M, Fontenille D (1999) Vectors of Chikungunya virus in Senegal: current data and transmission cycles. Am J Trop Med Hyg 60:281–286
Diaz LA, Ré V, Almirón WR, Farías A, Vázquez A, Sanchez-Seco MP, Aguilar J, Spinsanti L, Konigheim B, Visintin A, García J, Morales MA, Tenorio A, Contigiani M (2006) Genotype III Saint Louis encephalitis virus outbreak, Argentina, 2005. Emerg Infect Dis 12:1752–1754
Diaz LA, Llinás GA, Vázquez A, Tenorio A, Contigiani MS (2012) Silent circulation of St. Louis encephalitis virus prior to an encephalitis outbreak in Cordoba, Argentina (2005). PLoS Negl Trop Dis 6:e1489

Dick GW (1952) Zika virus. II. Pathogenicity and physical properties. Trans R Soc Trop Med Hyg 46:521–534

Dick GW, Kitchen SF, Haddow AJ (1952) Zika virus. I. Isolations and serological specificity. Trans R Soc Trop Med Hyg 46:509–520

Dobson AP (2005) What links bats to emerging infectious diseases? Science 310:628–629

Doherty RL, Gorman BM, Whitehead RH, Carley JG (1966) Studies of arthropod-borne virus infections in Queensland. V. Survey of antibodies to group A arboviruses in man and other animals. Aust J Exp Biol Med Sci 44:365–377

Dohm DJ, Rowton ED, Lawyer PG, O'Guinn M, Turell MJ (2000) Laboratory transmission of Rift Valley fever virus by *Phlebotomus duboscqi*, *Phlebotomus papatasi*, *Phlebotomus sergenti* and *Sergentomyia schwetzi* (Diptera: Psychodidae). J Med Entomol 37:435–438

Dorn PL, Melgar S, Rouzier V, Gutierrez A, Combe C, Rosales R, Rodas A, Kott S, Slavia D, Monroy CM (2003) The Chagas vector, *Triatoma dimidiata* (Hemiptera: Reduviidae), is panmictic within and among adjacent villages in Guatemala. J Med Entomol 40:436–440

Downs WG, Anderson CR, Spence L, Aitken THG, Greenhall AH (1963) Tacaribe virus, a new agent isolated from *Artibeus* bats and mosquitos in Trinidad, West Indies. Am J Trop Med Hyg 12:640–646

Duffy MR, Chen TH, Hancock WT, Powers AM, Kool JL, Lanciotti RS, Pretrick M, Marfel M, Holzbauer S, Dubray C, Guillaumot L, Griggs A, Bel M, Lambert AJ, Laven J, Kosoy O, Panella A, Biggerstaff BJ, Fischer M, Hayes EB (2009) Zika virus outbreak on Yap Island, federated states of Micronesia. N Engl J Med 360:2536–2543

Duval L, Robert V, Csorba G, Hassanin A, Randrianarivelojosia M, Walston J, Nhim T, Goodman SM, Ariey F (2007) Multiple host-switching of Haemosporidia parasites in bats. Malar J 29:157

Duval L, Mejean C, Maganga GD, Makanga BK, Mangama Koumba LB, Peirce MA, Ariey F, Bourgarel M (2012) The chiropteran haemosporidian *Polychromophilus melanipherus*: a worldwide species complex restricted to the family Miniopteridae. Infect Genet Evol 12:1558–1566

Eklund CM (1954) Mosquito-transmitted encephalitis viruses: a review of their insect and vertebrate hosts and the mechanisms for survival and dispersion. Exp Parasitol 3:285–305

Elliott RM (1997) Emerging viruses: the Bunyaviridae. Mol Med 3:572–577

Elliott RM, Blakqori G (2011) Molecular biology of orthobunyaviruses. In: Plyusnin A, Elliott RM (eds) Bunyaviridae, molecular and cell biology. Caister Academic, Norwich, pp 1–39

Ellis BR, Barrett DT (2008) The enigma of yellow fever in East Africa. Rev Med Virol 18:331–346

Ferro C, Boshell J, Moncayo AC, Gonzalez M, Ahumada ML, Kang W, Weaver SC (2003) Natural enzootic vectors of venezuelan equine encephalitis virus, Magdalena Valley, Colombia. Emerg Infect Dis 9:49–54

Fine DL, Roberts BA, Teehee ML, Terpening SJ, Kelly CLH, Raetz JL, Baker DC, Powers AM, Bowen RA (2007) Venezuelan equine encephalitis virus vaccine candidate (V3526) safety, immunogenicity and efficacy in horses. Vaccine 25:1868–1876

Flores FS, Diaz LA, Batallán GP, Almirón WR, Contigiani MS (2010) Vertical transmission of St. Louis encephalitis virus in *Culex quinquefasciatus* (Diptera: Culicidae) in Córdoba, Argentina. Vector Borne Zoonotic Dis 10:999–1002

Fontenille D, Traore-Lamizana M, Diallo M, Thonnon J, Digoutte JP, Zeller HG (1998) New vectors of Rift Valley fever in West Africa. Emerg Infect Dis 4:289–293

Fothergill LD, Dingle JH, Farber S, Connerly ML (1938) Human encephalitis caused by the virus of the eastern variety of equine encephalomyelitis. N Engl J Med 219:411

García L, Ortiz S, Osorio G, Torrico MC, Torrico F, Solari A (2012) Phylogenetic analysis of Bolivian bat trypanosomes of the subgenus *Schizotrypanum* based on cytochrome b sequence and minicircle analyses. PLoS One 7:e36578

Gardener RA, Molyneux DH (1988) *Polychromophilus murinus*: a malaria parasite of bats: life history and ultrastructural studies. Parasitology 96:591–605

Garmendia AE, Van Kruiningen HJ, French RA (2001) The West Nile virus: its recent emergence in North America. Microbes Infect 3:223–229

George S, Gourie-Devi M, Rao JA, Prasad SR, Pavri KM (1984) Isolation of West Nile virus from the brains of children who had died of encephalitis. Bull World Health Organ 62:879–882

Getting VA (1941) Equine encephalomyelitis in Massachusetts: an analysis of the 1938 outbreak. N Eng J Med 224:999–1006

Gillies MT, Coetzee M (1987) A supplement to the Anophelinae of Africa south of Sahara. Publ South Afr Inst Med Res 55:1–143

Goddard J, deShazo R (2009) Bed bugs (*Cimex lectularius*) and clinical consequences of their bites. JAMA 301:1358–1366

Gonzales JP, Georges AJ (1988) Bunyaviral fecers: Bunyamwera. Ilesha, Germiston, Bwamba, and Tataguine. In: Monath TP (ed) The arboviruses:epidemiology and Ecology, vol II. CRC, Boca Raton, FL, pp 87–98

Grady GF, Maxfield HK, Hildreth SW, Timperi RJ Jr, Gilfillan RF, Rosenau BJ, Francy DB, Calisher CH, Marcus LC, Madoff MA (1978) Eastern equine encephalitis in Massachusetts, 1957–1976. Am J Epidemiol 107:170–178

Grayson MA, Galindo P (1968) Epidemiologic studies of Venezuelan equine encephalitis virus in Almirante, Panama. Am J Epidemiol 88:80–96

Griffin DE (2001) Alphaviruses. In: Knipe DM, Howley PM (eds) Field virology. Lippincott Williams and Wilkins, Philadelphia, PA, pp 917–962

Gubler DJ (2007) The continuing spread of West Nile virus in the western hemisphere. Clin Infect Dis 45:1039–1046

Haddow AJ, Williams MC, Woodall JP, Simpson DIH, Goma LKH (1964) Twelve isolations of Zika virus from Aedes (Stegomyia) Africanus (Theobald) taken in and above a Uganda forest. Bull World Health Organ 31:57–69

Halpin K, Young PL, Field HE, Mackenzie JS (2000) Isolation of Hendra virus from pteropid bats: a natural reservoir of Hendra virus. J Gen Virol 81:1927–1932

Hamilton PB, Teixeira MGM, Stevens JR (2012a) The evolution of *Trypanosoma cruzi*: the 'bat seeding' hypothesis. Trends Parasitol 28:136–141

Hamilton PB, Cruickshank C, Stevens JR, Teixeira MMG, Mathews F (2012b) Parasites reveal movement of bats between the New and Old Worlds. Mol Phylogenet Evol 63:521–526

Hanson RP (1957) An epizootic of equine encephalomyelitis that occurred in Massachusetts in 1831. Am J Trop Med Hyg 6:858–862

Hardy JL, Apperson G, Asman SM, Reeves WC (1978) Selection of a strain of *Culex tarsalis* highly resistant to infection following ingestion of western equine encephalomyelitis virus. Am J Trop Med Hyg 27:313–321

Harley D, Ritchie S, Phillips D, van den Hurk A (2000) Mosquito isolates of Ross River virus from Cairns, Queensland, Australia. Am J Trop Med Hyg 62:561–565

Harley D, Sleigh A, Ritchie S (2001) Ross River virus transmission, infection and disease: a cross-disciplinary review. Clin Microbiol Rev 14:909–932

Hart KL, Keen D, Belle EA (1964) An outbreak of eastern equine encephalitis in Jamaica West Indies. I. Description of human cases. Am J Trop Med Hyg 13:331–341

Hayes CG (2006) West Nile virus: Uganda, 1937, to New York City, 1999. Ann N Y Acad Sci 951:25–37

Hayes EB (2009) Zika virus outside Africa. Emerg Infect Dis 15:1347–1350

Hayes RO, Daniels JB, Maxfield HK, Wheeler RE (1964) Field and laboratory studies on eastern encephalitis in warm- and coldblooded vertebrates. Am J Trop Med Hyg 13:595–606

Hemmersbach-Miller M, Parola P, Charrel RN, Durand JP, Brouqui P (2004) Sandfly fever due to Toscana virus: an emerging infection in southern France. Eur J Intern Med 15:316–317

Higgs S (2006) The 2005–2006 Chikungunya epidemic in the Indian Ocean. Vector Borne Zoonotic Dis 6:115–117

Hoshino K, Isawa H, Tsuda Y, Sawabe K, Kovbayashi M (2009) Isolation and characterization of a new insect *Flavivirus* from *Aedes albopictus* and *Aedes flavopictus* mosquitoes in Japan. Virology 391:119–129

Hu W, Tong S, Mengersen K, Oldenburg B, Dale P (2006) Mosquito species (Diptera: Culicidae) and the transmission of Ross River virus in Brisbane, Australia. J Med Entomol 43:375–381

Hubálek Z (2008) Mosquito-borne viruses in Europe. Parasitol Res 103:29–43

Hubálek Z, Halouzka J (1999) West Nile fever – a reemerging mosquito-borne viral disease in Europe. Emerg Infect Dis 5:643–650

Hutson AM, Mickleburgh SP, Racey PA (2001) Microchiropteran bats: global status survey and conservation active plan. IUCN/SSC Chiroptera Specialist Group. IUCN, Gland, Switzerland. 258 pp

Ikegami T (2012) Molecular biology and genetic diversity of Rift Valley fever virus. Antiviral Res 95:293–310

Ikegami T, Makino S (2011) The pathogenesis of Rift valley fever. Viruses 3:493–519

IUCN red list (2012) Information about different bat species. http://www.iucnredlist.org/

Iversson LB, Silva RAMS, Travassos da Rosa APA, Barros VLRS (1993) Circulation of eastern equine encephalitis, western equine encephalitis, Ilheus, Maguari and Tacaiuma viruses in equines of the Brazilian Pantanal, South America. Rev Inst Med Trop Sao Paulo 35:355–359

Jacob BG, Burkett-Cadena ND, Luvall JC, Parcak SH, McClure CJ, Estep LK, Hill GE, Cupp EW, Novak RJ, Unnasch TR (2010) Developing GIS-based eastern equine encephalitis vector-host models in Tuskegee, Alabama. Int J Health Geogr 9:12

Johnson HN (1964) Diseases derived from Wildlife. Proceedings of the 2nd vertebrate pest control conference, pp 138–142

Johnson KM, Martin DH (1974) Venezuelan equine encephalitis. Adv Vet Sci Comp Med 18:79–116

Johnson KM, Shelekov A, Peralta PH, Dammin GJ, Young NA (1968) Recovery of Venezuelan equine encephalitis virus in Panama. A fatal case in man. Am J Trop Med Hyg 17:432–440

Johnson BW, Cruz C, Felices V, Espinoza WR, Manock SR, Guevara C, Olson JG, Kochel TJ (2007) Ilheus virus isolate from a human, Ecuador. Emerg Infect Dis 13:956–958

Jones KE, Purvis A, MacLarnon A, Bininda-Emonds OR, Simmons NB (2002) A phylogenetic supertree of the bats (Mammalia: Chiroptera). Biol Rev Camb Philos Soc 77:223–259

Jöst H, Bialonski A, Storch V, Günther S, Becker N, Schmidt-Chanasit J (2010) Isolation and phylogenetic analysis of Sindbis viruses from mosquitoes in Germany. J Clin Microbiol 48:1900–1903

Kalayanarooj S, Vaughn DW, Nimmannitya S, Green S, Suntayakorn S, Kunentrasai N, Viramitrachai W, Ratanachu-eke S, Kiatpolpoj S, Innis BL, Rothman AL, Nisalak A, Ennis FA (1997) Early clinical and laboratory indicators of acute Dengue illness. J Infect Dis 176:313–321

Karabatsos, N. (ed) (1985). International catalogue of Arboviruses including certain other virus of vertebrates. 3rd ed., San Antonio, Texas. Am. Soc. Trop. Med. Hyg

Karstad LH, Hanson RP (1958) Infections in wildlife with the viruses of vesicular stomatitis and eastern equine encephalomyelitis. Trans N Am Wildl Conf 23:175–186

Kelser RA (1933) Mosquitoes as vectors of the virus of equine encephalomyelitis. J Am Vet Med Assoc 82:767

Kelser RA (1937) Transmission of the virus of equine encephalomyelitis by *Aedes taeniorhynchus*. Science 85:178

Kelser RA (1938) Transmission of the virus of equine encephalomyelitis by *Aedes taeniorhynchus*. J Am Vet Med Assoc 92:195

Knowlton GF, Rowe JA (1935) Handling mosquitoes on equine encephalomyelitis investigation. J Econ Entomol 28:824–829

Kramer LD, Scherer WF (1976) Vector competence of mosquitoes as a marker to distinguish Central American and Mexican epizootic from enzootic strains of Venezuelan encephalitis virus. Am J Trop Med Hyg 25:336–346

Krenn HW, Aspöck H (2010) Bau, Funktion und Evolution der Mundwerkzeuge blutsaugender Arthopoden. In: Aspöck H (ed) Krank durch Arthropoden. Denisia 30. Land Oberösterreich. Biologiezentum/Oberösterreichisches Landesmuseum, Linz, pp 81–108

Krenn HW, Aspöck H (2012) Form, function and evolution of the mouthparts of blood-feeding Arthropoda. Arthropod Struct Dev 41:101–118

Krishna MR, Reddy MK, Reddy SR (2006) Chikungunya outbreaks in Andhra Pradesh, South India. Curr Sci 91:570–571

Kumm HW (1932) Yellow fever transmission experiments with South American bats. Ann Trop Med Parasitol 26:207–213

Kurkela S, Manni T, Myllynen J, Vaheri A, Vapalahti O (2005) Clinical and laboratory manifestations of Sindbis virus infection: prospective study, Finland, 2002–2003. J Infect Dis 191:1820–1829

Kurkela S, Rätti O, Huhtamo E, Uzcátegui NY, Nuorti JP, Laakkonen J, Manni T, Helle P, Vaheri A, Vapalahti O (2008) Sindbis virus infection in resident birds, migratory birds, and humans, Finland. Emerg Infect Dis 14:41–47

Laemmert HW, Hughes T (1947) The virus of Ilheus encephalitis: isolation, serological specificity and transmission. J Immunol 55:61–67

Laine M, Luukkainen R, Toivanen A (2004) Sindbis viruses and other alphaviruses as cause of human arthritic disease. J Intern Med 256:457–471

Lambert AJ, Lanciotti RS (2008) Molecular characterization of medically important viruses of the genus *Orthobunyavirus*. J Gen Virol 89:2580–2585

Landau I, Rosin G, Miltgen F (1980) The genus *Polychromophilus* (Haemoproteidae, parasite of Microchiroptera). Ann Parasitol Hum Comp 55:13–32

Landau I, Chavatte JM, Karadjian G, Chabaud A, Beveridge I (2012) The haemosporidian parasites of bats with description of *Sprattiella alecto* gen. nov., sp. nov. Parasite 19:137–146

Lau SK, Woo PC, Li KS, Huang Y, Tsoi HW, Wong BH, Wong SS, Leung SY, Chan KH, Yuen KY (2005) Severe acute respiratory syndrome coronavirus-like virus in Chinese horseshoe bats. Proc Natl Acad Sci USA 102:14040–14045

LeDuc JW, Kahlon SS (2012) Bunyaviridae. In: Warrell D, Cox T, Firth J, Török E (eds) Oxford textbook of medicine: infection. Oxford University Press, Oxford, pp 182–185

Lee VH, Moore DL (1972) Vectors of the 1969 yellow fever epidemic on the Jos Plateau, Nigeria. Bull World Health Organ 46:669–673

Lee VH, Monath TP, Tomori O, Fagbami A, Wilson DC (1974) Arbovirus studies in Nupeko Forest, a possible natural focus of yellow fever in Nigeria. II. Entomological investigations and viruses isolated. Trans R Soc Trop Med Hyg 68:39–43

Lehane M (2005) The biology of blood-sucking in insects. Cambridge University Press, Cambridge, MA, 321 pp

Leroy EM, Kumulungui B, Pourrut X, Rouquet P, Hassanin A, Yaba P, Delicat A, Paweska JT, Gonzalez JP, Swanepoel R (2005) Fruit bats as reservoirs of Ebola virus. Nature 438:575–576

Letson GW, Bailey RE, Pearson J, Tsai TF (1993) Eastern equine encephalitis (EEE): a description of the 1989 outbreak, recent epidemiologic trends, and the association of rainfall with EEE occurrence. Am J Trop Med Hyg 49:677–685

Lewis DJ (1974) Biology of Phlebotomidae in relation to leishmaniasis. Annu Rev Entomol 19:363–384

Li L, Victoria JG, Wang C, Jones M, Fellers GM, Kunz TH, Delwart E (2010) Bat guano virome: predominance of dietary viruses from insects and plants plus novel mammalian viruses. J Virol 14:6955–6965

Löscher T, Burchard GD (2008) Tropenmedizin in Klinik und Praxis : mit Reise– und Migrationsmedizin. Georg Thieme Verlag, Stuttgart, 248 pp

Lundström JO, Pfeffer M (2010) Phylogeographic structure and evolutionary history of Sindbis virus. Vector Borne Zoonotic Dis 10:889–907

Lutwama JJ, Rwaguma EB, Nawanga PL, Mukuye A (2002) Isolations of Bwamba virus from south central Uganda and north eastern Tanzania. Afr Health Sci 2:24–28

Mackenzie JS, Williams DT (2008) The zoonotic flaviviruses of southern, south-eastern and eastern Asia, and Australiasia: the potential for emergent viruses. Zoonoses Public Health 56:338–356

Mackenzie JS, Lindsay MD, Coelen RJ, Broom AK, Hall RA, Smith DW (1994) Arboviruses causing human disease in the Australasian zoogeographic region. Arch Virol 136:447–467

Mackenzie JS, Gubler DJ, Petersen LR (2004) Emerging flaviviruses: the spread and resurgence of Japanese encephalitis, West Nile and Dengue viruses. Nat Med 10:98–109

Mackenzie JS, Williams DT, Smith DW (2006) Japanese encephalitis virus: the geographic distribution, incidence, and spread of a virus with a propensity to emerge in new areas. In: Tabor E (ed) Emerging viruses in human populations. Elsevier, Amsterdam, pp 201–268

Madsen DE, Knowlton GF (1935) Mosquito transmission of equine encephalomyelitis. J Am Vet Med Assoc 86:662

Madsen DE, Knowlton GF, Rowe JA (1936) Further studies on the transmission of equine encephalomyelitis by mosquitoes. J Am Vet Med Assoc 89:187

Maeda FY, Alves RM, Cortez C, Lima FM, Yoshida N (2011) Characterization of the infective properties of a new genetic group of *Trypanosoma cruzi* associated with bats. Acta Trop 120:231–237

Mahmood F, Crans WJ (1998) Effect of temperature on the development of *Culiseta melanura* (Diptera: Culicidae) and its impact on the amplification of eastern equine encephalitis virus in birds. J Med Entomol 35:1007–1012

Main AJ (1979a) Virologic and serologic survey for eastern equine encephalomyelitis and certain other viruses in colonial bats of New England. J Wildl Dis 15:455–464

Main AJ (1979b) Eastern equine encephalomyelitis virus in experimentally infected bats. J Wildl Dis 15:467–477

Marchette NJ, Garcia R, Rudnick A (1969) Isolation of Zika virus from *Aedes aegypti* mosquitoes in Malaysia. Am J Trop Med Hyg 18:411–415

Marcili A, Lima L, Cavazzana MR, Junqueira ACV, Veludo HH, Maia da Silva F, Campaner M, Paiva F, Nunes VLB, Teixeira MMG (2009) A new genotype of *Trypanosoma cruzi* associated with bats evidenced by phylogenetic analyses using SSU rDNA, cytochrome b and Histone H2B genes and genotyping based on ITS1 rDNA. Parasitology 136:641–655

Marinkelle CJ, Grose ES (1972) A review of bats as carriers of organisms which are capable of infecting man or domestic animals. Mitt Inst Colombo Aleman Invest Cient 6:31–51

Maroli M, Bigliocchi F, Khoury C (1994) Sandflies in Italy: observation on their distribution and methods for control. Parasitologia 36:251–264

McLean RG, Trevino HA, Sather GE (1979) Prevalence of selected zoonotic diseases in vertebrates from Haiti, 1972. J Wildl Dis 15:327–330

McMullan LK, Folk SM, Kelly AJ, MacNeil A, Goldsmith CS, Metcalfe MG, Batten BC, Albariño CG, Zaki SR, Rollin PE, Nicholson WL, Nichol ST (2012) A new Phlebovirus associated with severe febrile illness in Missouri. N Engl J Med 367:834–841

McMurray DN, Stroud J, Murphy JJ, Carlomagno MA, Greer DL (1982) Role of immunoglobulin classes in experimental histoplasmosis in bats. Dev Comp Immunol 6:557–567

Megali A, Yannic G, Christe P (2011) Disease in the dark: molecular characterization of *Polychromophilus murinus* in temperate zone bats revealed a worldwide distribution of this malaria-like disease. Mol Ecol 20:1039–1048

Mehlhorn H (ed) (2001) Encyclopedic reference of parasitology. Springer, Berlin, 676 pp

Merrill MH, Ten Broeck C (1935) The transmission of equine encephalomyelitis virus by *Aedes aegypti*. J Exp Med 62:687–695

Merrill MH, Lacaillade CW, Ten Broeck C (1934) Mosquito transmission of equine encephalitis. Science 80:251–252

Messenger SL, Rupprecht CE, Smith JS (2003) Bats, emerging virus infections, and the rabies paradigm. In: Kunz TH, Fenton MB (eds) Bat ecology. The University of Chicago Press, Chicago, pp 622–679

Meyer CG (2007) Tropenmedizin: Infektionskrankheiten. Hüthig Jehle Rehm GmbH, ecomed Medizin, 51 pp

Miller BR, Godsey MS, Crabtree MB, Savage HM, Al-Mazrao Y, Al-Jeffri MH, Abdoon A-M M, Al-Seghayer SM, Al-Shahrani AM, Ksiazek TG (2002) Isolation and genetic characterization of Rift Valley fever virus from *Aedes vexans arabiensis*, Kingdom of Saudi Arabia. Emerg Infect Dis 8:1492–1494

Modlmaier M, Kuhn R, Kaaden O-R, Pfeffer M (2002) Transmission studies of European Sindbis virus in the floodwater mosquito *Aedes vexans* (Diptera: Culicidae). Int J Med Microbiol 291:164–170

Mohrig W (1969) Die Culiciden Deutschlands. Parasitologische Schriftenreihe 18, Gustav Fischer Verlagm, Jena, 260 pp

Moore DL, Causey OR, Carey DE, Reddy S, Cooke AR, Akinkugbe FM, David-West TS, Kemp GE (1975) Arthropod-borne viral infections of man in Nigeria, 1964–1970. Ann Trop Med Parasitol 69:49–64

Morse RP, Bennish ML, Darras BT (1992) Eastern equine encephalitis presenting with a focal brain lesion. Pediatr Neurol 8:473–475

Murray KO, Mertens E, Desprès P (2010) West Nile virus and its emergence in the United States of America. Vet Res 41:67

Nabeshima T, Loan HTK, Inoue S, Sumiyoshi M, Haruta Y, Nga PT, Huoung VTQ, del Carmen PM, Hasebe F, Morita K (2009) Evidence of frequent introductions of Japanese encephalitis virus from south-east Asia and continental east Asia to Japan. J Gen Virol 90:827–832

Nalca A, Fellows PF, Whitehouse CA (2003) Vaccines and animal models for arboviral encephalitides. Antiviral Res 60:153–174

Netolitzky DJ, Schmaltz FL, Parker MD, Rayner GA, Fisher GR, Trent DW, Bader DE, Nagata LP (2000) Complete genomic RNA sequence of western equine encephalitis virus and expression of the structural genes. J Gen Virol 81:151–159

Norder H, Lundström JO, Kozuch O, Magnius LO (1996) Genetic relatedness of Sindbis virus strains from Europe, Middle East, and Africa. Virology 222:440–445

Novotny V, Basset Y, Miller SE, Weiblen GD, Bremer B, Cizek L, Drozd P (2002) Low host specificity of herbivorous insects in a tropical forest. Nature 416:841–844

Oelofsen MJ, van der Ryst E (1999) Could bats act as reservoir hosts for Rift Valley fever virus? Onderstepoort J Vet Res 66:51–54

Omatsu T, Watanabe S, Akashi H, Yoshikawa Y (2007) Biological characters of bats in relation to natural reservoir of emerging viruses. Comp Immunol Microbiol Infect Dis 30:357–374

Osborne JC, Rupprecht CE, Olson JG, Ksiazek TG, Rollin PE, Niezgoda M, Goldsmith CS, An US, Nichol ST (2003) Isolation of Kaeng Khoi virus from dead *Chaerephon plicata* bats in Cambodia. J Gen Virol 84:2685–2689

PAHO (2011) Preparedness and response for Chikungunya virus introduction in the Americas. http://new.paho.org/hq/index.php?option=com_docman&task=doc_download&gid=16984&Itemid=

Paul SD, Rajagopalan PK, Screenivasan MA (1970) Isolation of the West Nile virus from the frugivorous bat, *Rousettus leschenaultia*. Indian J Med Res 58:1–3

Pepin M, Bouloy M, Bird BH, Kemp A, Paweska J (2010) Rift Valley fever virus (Bunyaviridae: *Phlebovirus*): an update on pathogenesis, molecular epidemiology, vectors, diagnostics and prevention. Vet Res 41:61

Pilipski JD, Pilipski LM, Risley LS (2004) West Nile virus antibodies in bats from New Jersey and New York. J Wildl Dis 40:335–337

Pires DA, Gleiser RM (2010) Mosquito fauna inhabiting water bodies in the urban environment of Córdoba city, Argentina, following a St. Louis encephalitis outbreak. J Vector Ecol 35:401–409

Pittman PR, Makuch RS, Mangiafico JA, Cannon TL, Gibbs PH, Peters CJ (1996) Log-term duration of detectable neutralizing antibodies after administration of live-attenuated VEE vaccine and following booster vaccination with inactivated VEE vaccine. Vaccine 14:337–343

Platt KB, Mangiafico JA, Rocha OJ, Zaldivar ME, Mora J, Trueba G, Rowley WA (2000) Detection of Dengue virus neutralizing antibodies in bats from Costa Rica and Ecuador. J Med Entomol 37:965–967

Powers AM, Logue CH (2007) Changing patterns of chikungunya virus: re-emergence of a zoonotic arbovirus. J Gen Virol 88:2363–2377

Price JL (1978) Serological evidence of infection of Tacaribe virus and arboviruses in Trinidadian bats. Am J Trop Med Hyg 27:162–167

Rappole JH, Derrickson SR, Hubálek Z (2000) Migratory birds and spread of West Nile virus in the Western hemisphere. Emerg Infect Dis 6:319–328

Ravanini P, Huhtamo E, Ilaria V, Crobu MG, Nicosia AM, Servino L, Rivasi F, Allegrini S, Miglio U, Magri A, Minisini R, Vapalahti O, Boldorini R (2012) Japanese encephalitis virus RNA detected in *Culex pipiens* mosquitoes in Italy. Euro Surveill 17(28), pii = 20221

Ready P (2013) Biology of phlebotomine sand flies as vectors of disease agents. Annu Rev Entomol 58:227–250

Reagan RL, Rumbaugh H, Nelson H, Brueckner AL (1955) Effect of Zika virus and Bwamba virus in the cave bat (*Myotis lucifugus*). Trans Am Microsc Soc 74:77–79

Reinert JF, Harbach RE, Kitching IJ (2004) Phylogeny and classification of Aedini (Diptera: Culicidae), based on morphological characters of all life stages. Zool J Linn Soc 142:289–368

Reisen W (2003) Epidemiology of St. Louis encephalitis virus. Adv Virus Res 61:139–183

Reiskind MH, Wund MA (2009) Experimental assessment of the impacts of northern long-eared bats on ovipositing *Culex* (Diptera: Culicidae) Mosquitoes. J Med Entomol 49:1037–1043

Reiskind MH, Wund MA (2010) Bats & mosquitoes. Testing conventional wisdom. Bats 28:6–7

Reisler RB, Gibbs PH, Danner DK, Boudreau EF (2012) Immune interference in the setting of same-day administration of two similar inactivated alphavirus vaccines: eastern equine and western equine encephalitis. Vaccine 30:7271–7277

Reiter P, Fontenille D, Paupy C (2006) *Aedes albopictus* as an epidemic vector of chikungunya virus: another emerging problem? Lancet Infect Dis 6:463–464

Rezza G, Nicoletti L, Angelini R, Romi R, Finarelli AC, Panning M, Cordioli P, Fortuna C, Boros S, Magurano F, Silvi G, Angelini P, Dottori M, Ciufolini MG, Majori GC, Cassone A (2007) Infection with Chikungunya virus in Italy: an outbreak in a temperature region. Lancet 370:1840–1846

Rivas F, Diaz LA, Cardenas VM, Daza E, Bruzon L, Alcala A, De la Hoz O, Caceres FM, Aristizabal G, Martinez JW, Revelo D, De la Hoz F, Boshell J, Camacho T, Calderon L, Olano VA, Villarreal LI, Roselli D, Alvarez G, Ludwig G, Tsai T (1997) Epidemic Venezuelan equine encephalitis in La Guajira, Colombia, 1995. J Infect Dis 175:828–832

Rodrigues SG, Nunes MRT, Casseb SMM, Prazeres ASC, Rodrigues DSG, Silva MO, Cruz ACR, Tavares-Neto JC, Vasconcelos PFC (2010) Molecular epidemiology of Saint Louis encephalitis virus in the Brazilian Amazon: genetic divergence and dispersal. J Gen Virol 91:2420–2427

Rolle M, Mayr A (2007) Medizinische Mikrobiologie, Infektions- und Seuchenlehre: Für Tierärzte. Enke Verlag, Stuttgart, 627 pp

Romero JR, Newland JG (2003) Viral meningitis and encephalitis: traditional and emerging viral agents. Semin Pediatr Infect Dis 14:72–82

Rossi C, Rey O, Jenik P, Franze-Fernandez MT (1996) Immunological identification of Tacaribe virus proteins. Res Virol 147:203–211

Russell PK (1999) Vaccines in civilian defense against bioterrorism. Emerg Infect Dis 5:531–533

Russell RC (2002) Ross River virus: ecology and distribution. Annu Rev Entomol 47:1–31

Ryan PA, Martin L, Mackenzie JS, Kay BH (1997) Investigation of gray-headed flying foxes (*Pteropus poliochephalus*) (Megachiroptera: Pteropodidae) and mosquitoes in the ecology of Ross River virus in Australia. Am J Trop Med Hyg 57:476–482

Rydell J, Parker McNeill D, Eklöf J (2002) Capture success of little brown bats (*Myotis lucifugus*) feeding on mosquitoes. J Zool 256:379–381

Sailer R (1952) The bedbug. An old bedfellow that's still with us. Pest Control 10:22–24, 70, 72

Sallum MAM, Schultz TR, Wilkerson RC (2000) Phylogeny of Anophelinae (Diptera Culicidae) based on morphological characters. Ann Entomol Soc Am 93:745–775
Sam I-C, Path MRC, AbuBakar S (2006) Chikungunya virus infection. Med J Malaysia 61:264–270
Sambri V, Cavrini F, Rossini G, Pierro A, Landini MP (2008) The 2007 epidemic outbreak of Chikungunya virus infection in the Romagna region of Italy: a new perspective for the possible diffusion of tropical diseases in temperate areas. New Microbiol 31:303–304
Sanbonmatsu-Gámez S, Pérez-Ruiz M, Collao X, Sánchez-Seco MP, Morillas-Márquez F, De la Roas-Fraile M, Navarro-Marí JM, Tenorio A (2005) Toscana virus in Spain. Emerg Infect Dis 11:1701–1707
Sarkar SK, Chakravarty AK (1991) Analysis of immunocompetent cells in the bat, *Pteropus giganteus*: isolation and scanning electron microscopic characterization. Dev Comp Immunol 15:423–430
Schäfer M, Storch V, Kaiser A, Beck M, Becker N (1997) Dispersal behavior of adult snow melt mosquitoes in the Upper Rhine Valley, Germany. J Vector Ecol 22:1–5
Scherer WF, Campillo-Sainz C, de Mucha-Macias J, Dickerman RW, Chia CW, Zarate ML (1972) Ecologic studies of venezuelan encephalitis virus in southeastern Mexico. VII. Infection of man. Am J Trop Med Hyg 21:79–85
Schwartz O, Albert ML (2010) Biology and pathogenesis of Chikungunya virus. Nat Rev Microbiol 8:491–500
Scott TW (2001) Are bats really involved in Dengue virus transmission? J Med Entomol 38:771–772
Scott TW, Weaver SC (1989) Eastern equine encephalomyelitis virus: epidemiology and evolution of mosquito transmission. Adv Virus Res 37:277–328
Service MW (1993) Mosquitoes (Culicidae). In: Lane RP, Grosskey RW (eds) Medical insects and arachnids. Chapman & Hall, London, pp 120–240
Seymour C, Dickerman R, Martin MS (1978a) Venezuelan encephalitis virus infection in neotropical bats. I. Natural infection in a Guatemalan enzootic focus. Am J Trop Med Hyg 27:290–296
Seymour C, Dickerman R, Martin MS (1978b) Venezuelan encephalitis virus infection in neotropical bats. II. Experimental infections. Am J Trop Med Hyg 27:297–306
Simmons JS, Reynolds FHK, Cornell VH (1936) Transmission of the virus of equine encephalomyelitis through *Aedes albopictus*. Am J Trop Med Hyg 16:289–302
Simpson DI (1964) Zika virus infection in man. Trans R Soc Trop Med Hyg 58:335–338
Simpson DIH, O'Sullivan JP (1968) Studies on arbovirus on bats (Chiroptera) in East Africa. I.- Experimental infection of bats and virus transmission attempts in *Aedes* (*Stegomyia*) *aegypti* (Linnaeus). Ann Trop Med Parasitol 62:422–431
Smith DR, Adams AP, Kenney JL, Wang E, Weaver SC (2008) Venezuelan equine encephalitis virus in the mosquito vector *Aedes taeniorhynchus*: infection initiated by a small number of susceptible epithelial cells and a population bottleneck. Virology 372:176–186
Smith DW, Speers DJ, Mackenzie JS (2011) The viruses of Australia and the risk to tourists. Travel. Med Infect Dis 9:113–125
Smithburn KC, Mahaffy AF, Paul JH (1941) Bwamba fever and its causative agent. Am J Trop Med 21:75–90
Snodgrass RE (1959) The anatomical life of the mosquito. Smithsonian Misc Collect 139:1–85
Snow RW, Guerra CA, Noor AM, Myint HY, Hay SI (2005) The global distribution of clinical episodes of *Plasmodium falciparum* malaria. Nature 434:214–217
Soldan SS, González-Scarano F (2005) Emerging infectious diseases: the Bunyaviridae. J Neurovirol 11:412–423
Solomon T, Ni H, Beasley DWC, Ekkelenkamp M, Cardosa MJ, Barrett ADT (2003) Origin and evolution of Japanese encephalitis virus in Southeast Asia. J Virol 77:3091–3098
Spence L, Anderson CR, Downs WG (1962) Isolation of Ilheus virus from human beings in Trinidad, West Indies. Trans R Soc Trop Med Hyg 56:504–509

Spinsanti L, Basquiera AL, Bulacio S, Somale V, Kim SCH, Ré V, Rabbat D, Zárate A, Zlocowski JC, Mayor CR, Contigiani M, Palacio S (2003) St. Louis encephalitis in Argentina: the first case reported in the last seventeen years. Emerg Infect Dis 9:271–273

Srihongse S, Johnson CM (1967) Isolation of Ilheus from man in Panama. Am J Trop Med Hyg 16:516–518

Strauss JH, Strauss EG (1994) The alphaviruses: gene expression, replication, and evolution. Microbiol Rev 58:491–562

Sudia WD, Newhouse VF, Henderson BE (1971) Experimental infection of horses with three strains of Venezuelan equine encephalomyelitis virus II. Experimental vector studies. Am J Epidemiol 93:206–211

Sulkin SE, Allen R, Miura T, Toyokawa K (1970) Studies of arthropod-borne virus infections in Chiroptera. VI Isolation of Japanese B encephalitis virus from naturally infected bats. Am J Trop Med Hyg 19:77–87

Tajima S, Takasaki T, Matsuno S, Nakayama M, Kurane I (2005) Genetic characterization of Yokose virus, a *Flavivirus* isolated from the bat in Japan. Virology 332:38–44

Teeling EC, Springer MS, Madsen O, Bates P, O'Brien SJ, Murphy WJ (2005) A molecular phylogeny for bats illuminates biogeography and the fossil record. Science 307:580–584

Tesh RB (1982) Arthritides caused by mosquito-borne viruses. Annu Rev Med 33:31–40

Tesh RB, Jahrling PB, Salas R, Shope RE (1994) Description of Guanarito virus (Arenaviridae: *Arenavirus*), the etiologic agent of Venezuelan hemorrhagic fever. Am J Trop Med Hyg 50:452–459

Thomas SM, Fischer D, Fleischmann S, Bittner T, Beierkuhnlein C (2011) Risk assessment of Dengue virus amplification in Europe based on spatio-temporal high resolution climate change projectations. Erdkunde 65:137–150

Tiawsirisup S (2011) Chikungunya virus in Thailand; an update. Thai J Vet Med 41:133–134

Tiawsisirsup S, Junpee A, Nuchprayoon S (2012) Mosquito distribution and Japanese encephalitis virus infection in a bat cave and its surrounding area in Lopburi province, Central Thailand. Thai J Vet Med 42:43–49

Tikasingh ES, Ardoin P, Williams MC (1974) First isolation of Catú virus from a human in Trinidad. Trop Geogr Med 26:414–416

Tsetsarkin KA, Vanlandingham DL, McGee CE, Higgs S (2007) A single mutation in Chikungunya virus affects vector specificity and epidemic potential. PLoS Pathog 3:e201

Ubico SR, McLean RG (1995) Serologic survey of neotropical bats in Guatemala for virus antibodies. J Wildl Dis 31:1–9

Usinger RL (1966) Monograph of Cimicidae. Entomological Society of America, Washington, DC, 585 pp

Valassina M, Cusi MG, Valensin PE (2003) A Mediterranean arbovirus: the Toscana virus. J Neurovirol 9:577–583

Van der Hurk AF, Ritchie SA, Mackenzie JS (2009a) Ecology and geographical expansion of Japanese encephalitis virus. Annu Rev Entomol 54:17–35

Van der Hurk AF, Smith CS, Field HE, Smith IL, Northill JA, Taylor CT, Jansen CC, Smith GA, Mackenzie JS (2009b) Transmission of Japanese encephalitis virus from the black flying fox, Pteropus alecto, to Culex annulirostris mosquitoes, despite the absence of detectable viremia. Am J Trop Med Hyg 81:457–462

Venegas EA, Aguilar PV, Cruz C, Guevara C, Kochel TJ, Vargas J, Halsey ES (2012) Ilheus virus infection in human, Bolivia. Emerg Infect Dis 18:516–518

Wang J-L, Pan X-L, Zhang H-L, Fu S-H, Wang H-Y, Tang Q, Wang L-F, Liang G-D (2009) Japanes encephalitis viruses from bats in Yunnan, China. Emerg Infect Dis 15:939–942

Weaver SC, Bellew LA, Rico-Hesse R (1992) Phylogenetic analysis of alphaviruses in the Venezuelan equine encephailitis complex and identification of the source of epizootic viruses. Virology 191:282–290

Weaver SC, Ferro C, Barrera R, Boshell J, Navarro JC (2004) Venezuelan equine encephalitis. Annu Rev Entomol 49:141–174

Weidmann M, Rudaz V, Nunes MRT, Vasconcelos PFC, Hufert FT (2003) Rapid detection of human pathogenic orthobunyaviruses. J Clin Microbiol 41:3299–3305

Weinstein P, Judge D, Carver S (2011) Biological and cultural coevolution and emerging infection disease: Ross River virus in Australia. Med Hypotheses 76:893–896

Whitley RJ, Gnann JW (2002) Viral encephalitis: familiar infections and emerging pathogens. Lancet 359:507–513

WHO (2000) Report on global surveillance of epidemic-prone infectious diseases. WHO/CDS/CSR/ISR/20001, pp 1–15

WHO (2004) The vector-borne human infections of Europe – their distribution and burden on public health. http://www.euro.who.int/__data/assets/pdf_file/0008/98765/e82481.pdf

WHO (2009) Dengue: guidelines for diagnosis, treatment, prevention and control. http://www.who.int/rpc/guidelines/9789241547871/en/

WHO (2011) West Nile virus, fact sheet N° 354. http://www.who.int/mediacentre/factsheets/fs354/en/index.html

WHO (2012) Global alert and response (GAR), Rift Valley fever in Mauretania. http://www.who.int/csr/don/2012_11_01/en/index.html

WHO (2013a) Global alert and response (GAR), Rift Valley fever. http://www.who.int/csr/disease/riftvalleyfev/en/

WHO (2013b) Rift Valley fever, fact sheet N° 207. http://www.who.int/mediacentre/factsheets/fs207/en/

Williams MC, Simpson DIH, Shepherd RC (1964) Bats and arbovirus in east Africa. Nature 203:670

Williams JE, Imlarp S, Top FH Jr, Cavanaugh DC, Russell PK (1976) Kaeng Khoi virus from naturally infected bedbugs (Cimicidae) and immature free-tailed bats. Bull World Health Organ 53:365–369

Witsenburg F, Salamin N, Christe P (2012) The evolutionary host switches of *Polychromophilus*: a multi-gene phylogeny of the bat malaria genus suggests a second invasion of mammals by a haemosporidian parasite. Malar J 11:53

Wong S, Lau S, Woo P, K-Ym Y (2007) Bat as a continuing source of emerging infections in humans. Rev Med Virol 17:67–91

Woods CW, Karpati AM, Grein T, McCarthy N, Gaturuku P, Muchiri E, Dunster L, Henderson A, Khan AS, Swanepoel R, Bonmarin I, Martin L, Mann P, Smoak BL, Ryan M, Ksiazek TG, Arthur RR, Ndikuyeze A, Agata NN, Peters CJ (2002) An outbreak of Rift Valley fever in northeastern Kenya, 1997–98. Emerg Infect Dis 8:138–144

Zacks MA, Paessler S (2010) Encephalitic alphaviruses. Vet Microbiol 140:281–286

Zeledón R, Rabinovich JE (1981) Chagas' disease: an ecological appraisal with special emphasis on its insect vectors. Annu Rev Entomol 26:101–133

Zeller HG, Schuffenecker I (2004) West Nile virus: an overview of its spread in Europe and the Mediterranean basin in contrast to its spread in the Americas. Eur J Clin Microbiol Infect Dis 23:147–156

Chapter 4
Bat Endoparasites: A UK Perspective

Jennifer S. Lord and Darren R. Brooks

Abstract Studies have shown that bats are infected with a rich community of endoparasites. However, detailed investigations are lacking; not least because of the challenges of working with hosts that are protected by legislation. Below, we review the status of bat endoparasite studies in the UK, giving due consideration to a significant body of classical parasitological investigations on haematozoa (trypanosomes, the piroplasm *Babesia vesperuginis* and the haemosporidian *Polychromophilus murinus*) carried out in the mid-1980s on almost 500 hosts and encompassing 12 of the 17 bat species known to breed in the UK. Of these parasites, only *B. vesperuginis*-infected bats showed any adverse health impacts, including elevated reticulocyte and white blood cell counts, reduced haemoglobin levels, haemoglobinuria and splenomegaly. More recently, molecular-based analyses of UK bat haematozoa have contributed not only to enriching survey data but also importantly to a wider understanding of evolutionary relationships amongst parasites, which in turn has provided insight into historic movements of the hosts. We also discuss gastrointestinal parasite infections and highlight the lack of published studies on UK bat coccidians and helminths. As such, morphological and molecular analyses carried out in our laboratory, on a population of pipistrelle bats in South Lancashire and Greater Manchester, are providing baseline data on these infections in UK bats. With regard to helminths, we find that pipistrelle bats are commonly infected with digenean trematodes (prevalence = 76 %; mean abundance = 48.2 ± 7), particularly lecithodendriids (e.g. *Lecithodendrium linstowi*). Moreover, helminth infections were significantly more aggregated and also less abundant in male bats compared to females, an interesting and perhaps

J.S. Lord
Department of Biology and Emerging Pathogens Institute, University of Florida, Gainesville, FL 32611, USA

D.R. Brooks (✉)
Ecosystems and Environment Research Centre, School of Environment and Life Sciences, University of Salford, Salford, Greater Manchester M5 4WT, UK
e-mail: d.r.brooks@salford.ac.uk

S. Klimpel and H. Mehlhorn (eds.), *Bats (Chiroptera) as Vectors of Diseases and Parasites*, Parasitology Research Monographs 5, DOI 10.1007/978-3-642-39333-4_4,

surprising parasite response to the host sex hormones. DNA sequencing of the 28S rRNA gene of representative lecithodendriid specimens has offered new insight into evolutionary relationships amongst the Lecithodendriidae, specifically, separating a controversial clade between *L. linstowi* and *Prosthodendrium hurkovaae*. Finally, we highlight recent work that utilises PCR-based detection to implicate bats as potentially important reservoir hosts of the apicomplexan *Toxoplasma gondii* (prevalence = 10 %).

With respect to bat health and disease, although endoparasite infections have been linked to specific pathologies (e.g. splenomegaly), compelling recent evidence suggests that the overall rate of severe disease is low (10 %) and that the mortality rate, due primarily to endoparasite infection, is minimal (0.5 %). As such, the relationship between bats and their endoparasites is probably evolutionarily ancient and reflective of a healthy ecosystem.

Importantly, bat endoparasite research, both in the UK and elsewhere, is contributing to a greater understanding of parasite biodiversity and wildlife health and disease. Significant challenges remain; not least understanding the community ecology of bat parasites and how these infections impact upon other important infectious diseases of bats (viral, bacterial and fungal). Progress will be dependent upon multidisciplinary approaches that encompass the expertise of many individuals including parasitologists, bat biologists, veterinarians and volunteer members of bat conservation groups.

Keywords Chiroptera • Endoparasite • Protozoa • Haematozoa • Trypanosome • Piroplasm • Haemosporidia • Helminth • Trematode

4.1 Introduction

In the UK, like many other areas of the world, bats have undergone widespread population decline and as a consequence, all species are protected by national and international legislation. There are 17 species of bat known to breed in the UK, all of which are insectivorous and belong to the families Rhinolophidae and Vespertilionidae. Although this number is a small proportion of the species known worldwide, it is of national significance since it represents approximately one-quarter of all UK mammal species. Like elsewhere, anthropogenic factors have had the most significant impact upon bat populations; this is particularly manifest in habitat loss, environmental change and pollution, including widespread use of insecticides (Harris et al. 1995). As a consequence of the rapid population declines that have occurred, two bat species are of particular concern with respect to their near threatened status on the IUCN Red List of Threatened Species: *Myotis bechsteinii* and *Barbastella barbastellus* (Hutson et al. 2008a, b). Moreover, the mouse-eared bat, *Myotis myotis*, is categorised as the most recent mammal extinction in the UK and some other breeding species (e.g. *Rhinolophus ferrumequinum*, *Plecotus austriacus* and *Nyctalus leisleri*) are infrequently encountered and hence considered rare in the UK (Bat Conservation Trust 2011; Harris et al. 1995).

Most studies of UK bats are in relation to understanding bat ecology and conservation (for example Hale et al. 2012; Rivers et al. 2005) and also the role of bats in acting as reservoirs of zoonotic disease, specifically lyssavirus type 2, the agent responsible for the sole reported fatal case of UK rabies transmission from a bat, *M. daubentonii*, to a human (Fooks et al. 2003; Harris et al. 2009). Given the extensive literature focussed on bat viruses (Calisher et al. 2006; Wang et al. 2011), the purpose of this chapter is to highlight other infectious agents of UK bats; specifically their endoparasites. Indeed, given that bats are the second most speciose mammalian order, bat parasitology remains vastly under-represented in the literature. This no doubt reflects the difficulties associated with working with protected species and also health and safety considerations. Nonetheless, there has been a sporadic series of bat parasite research efforts in the UK, that in conjunction with studies elsewhere, have made significant contributions to our current understanding of the biology of these understudied parasitic organisms.

The impact of White-nose syndrome upon North American bat populations (Frick et al. 2010) has highlighted the need for continued research efforts to further our understanding of bat health and disease. Bat parasite studies will contribute to this knowledge base and more broadly, the growing field of wildlife parasitology. This in turn will inform and shape opinion on diverse areas; not least current understanding of zoonotic diseases, phylogenetic relationships and parasite community ecology.

4.2 Haematozoa

It has been recognised since the late 1800s that bats harbour blood-borne parasites (Dionisi 1898, 1899). In the early part of the twentieth century, blood smears analysed from limited numbers of UK bats confirmed the presence of trypanosomes, piroplasms and the malaria-like parasite *Polychromophilus murinus* (Coles 1914; Petrie 1905) (Table 4.1). Although further studies were carried out in other parts of the world, no additional reports of UK bat haematozoa were documented until a series of studies in the 1970s–1980s (Table 4.1). The most extensive work during this period was carried out by Gardner and Molyneux (Table 4.2); blood smears from almost 500 bats, representing 12 species, captured across southern and central England (Suffolk, Hampshire, Sussex, Kent, Wiltshire, Dorset, Cambridgeshire, Norfolk, East Anglia, Northamptonshire and Staffordshire) and Scotland (Aberdeenshire and Inverness-shire), were examined and attempts were made to identify the intermediate hosts most likely responsible for parasite transmission (Gardner 1986; Gardner and Molyneux 1987, 1988a, b, c; Gardner et al. 1987). Below, we provide further details of Gardner and Molyneux's significant contributions to bat parasitology and highlight more recent, molecular-based studies of the haematozoa of UK bats, including work in our own laboratory on *P. pipistrellus* and *P. pygmaeus* bats acquired from South Lancashire and Greater Manchester.

Table 4.1 Summary of UK bat haematozoa studies

Location	Host species	Numbers analysed	Parasites identified	References
Hertfordshire	*P. pipistrellus*	8	3 infected with *T. vespertilionis*[a]	Petrie (1905)
	N. noctula	1	0	
Berkshire	*P. pipistrellus*	13	3 infected with *T. vespertilionis*; 7 infected with *Achromaticus* (=*Babesia*) *vesperuginis*	Coles (1914)
Dorset	*P. pipistrellus*	7	1 infected with *P. murinus*	
	N. noctula	3	0	
	P. auritus	2	0	
Norfolk	*P. pipistrellus*	10	3 infected with *T. vespertilionis*[b]; 1 infected with *P. murinus*; 1 infected with *B. vesperuginis*	Molyneux and Baffort (1971)
Cambridgeshire	*N. noctula*	5	2 infected with *T. vespertilionis*	Baker and Thompson (1971)
East Anglia	*P. pipistrellus*	8	5 infected with *T. dionisii*; 2 infected with *B. vesperuginis*	Baker et al. (1972)
Norfolk	*P. pipistrellus*	21	18 infected with *T. vespertilionis*; 3 infected with *T. incertum*	Baker (1973, 1974)
Not specified	*M. daubentonii*	Not specified	*P. murinus*	Hutson (1984)
Cornwall	*Pipistrellus* sp.	36	6 infected with *B. vesperuginis*; 1 infected with *T. dionisii*	Concannon et al. (2005)
	B. barbastellus	1	0	
	P. auritus	15	0	
	R. hipposideros	3	0	
	M. mystacinus	2	0	
	M. daubentonii	1	0	
	N. noctula	1	0	
	M. nattereri	1	0	
South Lancashire/ Greater Manchester	*P. pipistrellus*	93	15 infected with *B. vesperuginis*; 28 infected with *Trypanosoma* spp.[c];	Lord (2010)

(continued)

Table 4.1 (continued)

Location	Host species	Numbers analysed	Parasites identified	References
			5 co-infected with *B. vesperuginis* and *Trypanosoma* spp.[c]	
	P. pygmaeus	6	1 infected with *Trypanosoma* spp.[c]; 3 co-infected with *B. vesperuginis* and *Trypanosoma* spp.[c]	
	M. mystacinus	1	0	
Oxfordshire	*P. pygmaeus*	13	7 infections with *T. dionisii*	Hamilton et al. (2012b)
North Somerset	*Nyctalus noctula*	8	3 infections with *T. dionisii* B (New 2) and one of these co-infected with *T. vespertilionis*	
Wiltshire	*M. mystacinus*	6	2 infections with *T. dionisii* B (New 1)	
	E. serotinus	13	3 infections with *T. dionisii* B (New 1)	
Oxfordshire/ Wiltshire/ North Somerset	*B. barbastellus, M. bechsteinii, P. auritus, M. brandtii, M. daubentonii, M. nattereri, R. ferrumequinum, R. hipposideros*	98	0	

For studies published between 1905 and 1974, *P. pipistrellus* specimens should be treated as potentially also including *P. pygmaeus*. Gardner and Molyneux's data is summarised in Table 4.2
[a]Should be treated as *Trypanosoma (Schizotrypanum)* sp.
[b]Most likely to be *T. dionisii*
[c]Thirty of these trypanosome infections were diagnosed as *T. dionisii*

Table 4.2 Summary of Gardner and Molyneux's bat haematozoa studies

Host	Number examined	*Trypanosoma (Schizotrypanum)* infections	*Trypanosoma (Megatrypanum)* infections	*Babesia vesperuginis* infections	*Polychromophilus murinus* infections	Prevalence of haematozoa (%)
Pipistrellus pipistrellus	206[a,b]	73	33[c]	19[d]	0	25
Plecotus auritus	51	0	0	0	0	0
Nyctalus noctula	27	1	0	0	0	4
N. leisleri	8	5	0	0	0	63
Eptesicus serotinus	15	5[e]	0	0	0	33
Myotis daubentonii	108	0	0	0	27	25
M. nattereri	13	0	0	0	0	0
M. brandtii	6	1	0	0	0	17
M. mystacinus	11	0	0	1	0	9
M. bechsteinii	1	0	0	0	0	0
Rhinolophus ferrumequinum	38	0	0	0	0	0
R. hipposideros	7	0	0	0	0	0
Total population	491	85 (17 % prevalence)	33 (7 % prevalence)	20 (4 % prevalence)	27 (5 % prevalence)	

P. pipistrellus specimens should be treated as potentially also including *P. pygmaeus*. Data extracted from Gardner (1986)
[a,b]Includes five *Schizotrypanum*/*Megatrypanum* and six *Schizotrypanum*/*B. vesperuginis* co-infections
[c]26 infections identified in bats from Aberdeenshire
[d]17 infections identified in bats from Cambridgeshire
[e]All identified as *T. dionisii*

4.2.1 *Trypanosomes*

Trypanosomes are flagellated, kinetoplastid parasites that infect and cause diseases in many organisms including humans. The biology of the bat trypanosomes is less well understood; however, in comparison to other bat endoparasites, some species of bat trypanosomes are well studied and Molyneux provides a comprehensive review (Molyneux 1992). Blood smear analyses have confirmed the presence of two sub-genera of trypanosomes in UK bats: the *Schizotrypanum* and the *Megatrypanum*. Species of bat *Schizotrypanum* are known to encyst in various organs and tissues including the heart and skeletal muscles (Molyneux 1992). However, the significance of this upon the health of the host requires further investigations.

4.2.1.1 *Schizotrypanum* in UK Bats

The *Schizotrypanum* type species is the human infective parasite *Trypanosoma cruzi*, which is also known to infect some South American bats and be transmitted congenitally (Anez et al. 2009). However, the role of bats as reservoir species for transmission of *T. cruzi* to humans and other animals requires further exploration (Ramsey et al. 2012).

Trypanosoma (S.) vespertilionis and *T. (S.) dionisii* are known to be associated with European bats, and these parasites were documented by Gardner and Molyneux (1988b) in five species of UK bat: *Pipistrellus pipistrellus*, *Nyctalus leisleri*, *N. noctula*, *Eptesicus serotinus* and *Myotis brandtii* (Table 4.2). Utilising morphological assessment of in vitro cultivated parasites isolated from bat blood and also DNA analysis, the predominant trypanosome was noted as *T. (S.) dionisii*. Total *Schizotrypanum* prevalence across the UK was recorded as 17 % (85/491), though there was significant regional variation, for example 0 % (0/10) in Inverness-shire, Scotland, and 33 % (29/88) in Suffolk (Gardner 1986). *P. pipistrellus* was the most commonly infected bat species (35 %, 73/206). Given that the soprano pipistrelle has only relatively recently been recognised as a separate species (Barratt et al. 1997), it is probable that some of these infection records may be associated with *P. pygmaeus*. Gardner and Molyneux also qualified their infection data as likely to be an under-representation of *Schizotrypanum* infection in UK bats, given that the analysis was based upon extraction of peripheral blood taken from the uropatagial vein in the tail membrane (Gardner and Molyneux 1988b). Given the numbers of bats examined, the absence of infections noted in *Plecotus auritus* (0/51), *M. daubentonii* (0/108) and *Rhinolophus ferrumequinum* (0/38) may be of significance and reflect the preferred host range of the intermediate vector host (Gardner 1986).

More recent work, based upon nested PCR amplification of a fragment of the *Trypanosoma* spp. 18S rRNA gene from DNA prepared from bat heart, demonstrated that of 60 bats acquired from Cornwall, only one, a pipistrelle specimen, was infected with *Schizotrypanum* (Concannon et al. 2005) (Table 4.1).

DNA sequence analysis identified the source of this single infection as *T. (S.) dionisii* (Concannon et al. 2005). Similar PCR-based work in our own laboratory, on a sample of 99 pipistrelle bats acquired across South Lancashire and Greater Manchester, showed that 36 % were infected with *Schizotrypanum* (Table 4.1). Consistent with the Gardner and Molyneux study, the majority of these infections were speciated as *T. (S.) dionisii* (Lord 2010).

Most recently, a molecular-based study of 13 bat species in southern England reported *Schizotrypanum* prevalence as 10 % (Table 4.1) (Hamilton et al. 2012b). With the exception of one *N. noctula* bat with a mixed *T. (S.) dionisii*, *T. (S.) vespertilionis* infection, all remaining infections (*P. pipistrellus, P. pygmaeus, E. serotinus* and *M. mystacinus*) were attributable to *T. (S.) dionisii* (Hamilton et al. 2012b). As noted by Gardner and Molyneux (Gardner 1986), no *Schizotrypanum* infections were present in *M. daubentonii* (0/26), *R. ferrumequinum* (0/15) and *P. auritus* (0/12).

The reported regional differences in prevalence data most likely reflect contrasting dynamics between local populations of bats and the intermediate host, the bat bug *Cimex pipistrelli*. The latter was suspected as the vector following identification of infective metacyclic trypanosomes from *C. pipistrelli* sampled from several UK roosting sites (Gardner and Molyneux 1988b). Subsequent laboratory analyses, including assessment of the infectivity of *Schizotrypanum* to *C. pipistrelli*, have provided further insight into the developmental cycle of these trypanosomes within the vector (Gardner and Molyneux 1988b).

With increasing improvements to genetic techniques, it is becoming feasible to assess the relatedness of parasite isolates and use this data as an indicator of potential host movements. Interestingly, fluorescent fragment length bar coding, a technique that analyses multiple ribosomal RNA gene regions, has allowed identification of two new *T. (S.) dionisii* genotypes in *E. serotinus* and *M. mystacinus* bats in southern England (Table 4.1) (Hamilton et al. 2012b). DNA sequencing and phylogenetic analyses, based on 18S rRNA and gGAPDH gene sequences, revealed that these were two new strains of *T. (S.) dionisii*. Most intriguingly, the new strains appeared closely related to *T. (S.) dionisii* parasites from Brazil. This type of analysis, in addition to the study of new species of *Schizotrypanum* isolated from bats in other regions of the world (Lima et al. 2012), is providing compelling evidence for a 'bat seeding' hypothesis to explain the evolutionary history of *T. cruzi* (Hamilton et al. 2012a). Additional phylogeographic analyses of bat trypanosomes should not only continue to test this hypothesis but also facilitate detailed modelling of historic bat movements and so provide insight into their current distributions. Moreover, this should assist a wider understanding of patterns of current and emerging zoonotic disease associated with bats.

4.2.1.2 *Megatrypanum* in UK Bats

Trypanosoma incertum was the first *Megatrypanum*, or large trypanosome, to be recorded in a European bat following blood smear analysis of a pipistrelle sampled from East Anglia (Baker 1973). The extensive survey of UK bats by Gardner and Molyneux showed that *T. incertum* was highly host restricted, being recorded in 16 % (33/206) of pipistrelle bats but absent from all other species examined (Table 4.2) (Gardner 1986; Gardner and Molyneux 1988c). Closer inspection of this survey indicates that many of these parasites were recorded from bats sampled from a single roost in Aberdeenshire, Scotland (26/78). Small numbers of mixed infections with *Schizotrypanum* were also recorded (Gardner 1986).

Compared to *Schizotrypanum* parasites of bats, much less is known about the biology of *T. incertum*. However, laboratory studies have established that the bat bug *C. pipistrelli* is readily infected with *T. incertum* and that full development to infective stages can occur over the course of 3 days (Gardner and Molyneux 1988c).

There is an absence of more recent data on *T. incertum* in UK bats, or indeed bats from elsewhere. This most likely reflects the difficulties of sampling and also, the lack of molecular tools available to survey for *T. incertum*, as opposed to the parasite becoming increasingly scarce. To this end, a recent phylogeographical study of ruminant *Megatrypanum* parasites (Garcia et al. 2011) may assist in the design of suitable PCR primers for the detection of *T. incertum* infection in bats.

4.2.2 Piroplasms *in UK Bats*

A piroplasm, speciated as *Babesia vesperuginis* based upon host range, intra-erythrocytic location and morphology, has been recorded in UK pipistrelle bats following blood smear analyses (Tables 4.1 and 4.2). Unfortunately, there is no representative isolate of *B. vesperuginis* and hence a genetic record of this parasite is currently unavailable. The extensive survey conducted Gardner and Molyneux (1987) recorded overall *B. vesperuginis* prevalence as 4 % (20/491) (Table 4.2). With the exception of one *M. mystacinus* sampled from a roost in Wiltshire, all remaining *B. vesperuginis* infections were in pipistrelle bats (9 %, 19/206); the majority (17/19) of these being from sites in Cambridgeshire (Gardner 1986). Ectoparasite examination of piroplasm infected pipistrelles showed that the argasid tick *Argas vespertilionis* was frequently present and hence most likely to be the vector responsible for transmission (Gardner and Molyneux 1987). Further investigations are required to provide absolute confirmation of the intermediate host and describe parasite development within it.

More recently, a molecular approach based upon PCR amplification of 18S rRNA gene fragments specific to the Piroplasmida was applied to heart DNA extracted from bats sampled from Cornwall (Concannon et al. 2005). This study recorded an overall piroplasm prevalence of 10 % (6/60) and noted that all positive infections were in pipistrelle bats (6/36). DNA sequencing and phylogenetic analysis of the pipistrelle-associated piroplasm 18S rRNA gene fragment showed that it clustered specifically with other members of the Babesiidae family; the closest evolutionary relationships were to piroplasms from deer, though the bootstrap support was lacking significance (Concannon et al. 2005). Nonetheless, Concannon and colleagues propose sound arguments that this pipistrelle-associated piroplasm 18S rRNA gene sequence (GenBank: AJ871610) is the first unequivocal identifier for *B. vesperuginis* (Concannon et al. 2005).

Since piroplasms were described in bats, they have been suspected of having an adverse health impact upon the host (Dionisi 1898). Blood film examination of *B. vesperuginis*-infected pipistrelles acquired by Gardner and Molyneux showed that they had significantly higher reticulocyte counts than non-infected bats (Gardner and Molyneux 1987). Further blood and pathological analyses, on wild pipistrelles naturally infected with piroplasms and also experimental infections of *B. vesperuginis*, confirmed that these parasites are capable of eliciting detrimental health impacts upon the host: significantly elevated reticulocyte and white blood cell counts, significantly lowered haemoglobin levels, haemoglobinuria and splenomegaly were described (Gardner 1986; Gardner and Molyneux 1987). Peak parasitaemias were also noted as coinciding with some bats displaying a reluctance to fly and being incapable of maintaining their flight (Gardner and Molyneux 1987). Gardner and Molyneux conclude that babesiosis in wild bats will prevent effective foraging behaviour and possibly impact upon the ability of the spleen to function effectively in blood regulation during daily torpor and hibernation (Gardner and Molyneux 1987).

In other work, a veterinary pathology examination of 245 UK bats, carried out between 1990 and 1994, showed that 6 % of the specimens had splenomegaly; moreover, one of these bats was recorded as dying from babesiosis (Simpson 2000). In contrast to the aforementioned studies, the analysis of parasite infections in Cornish bats reported that although average spleen size in pipistrelles infected with *B. vesperuginis* was greater than that of non-infected specimens, this difference was not statistically significant (Concannon et al. 2005). This may have been a consequence of too few infected specimens in the dataset to make the comparison meaningful, and/or possibly, the effects of long-term storage of bat carcasses prior to sample analysis.

In our laboratory, PCR-based detection of piroplasm infection in bats sampled across South Lancashire and Greater Manchester showed that 23 % of pipistrelles were infected with *B. vesperuginis* (Lord 2010). As reported in other studies, we also noted significant splenomegaly associated with these piroplasm infections; as a percentage of body weight, spleens in infected individuals were approximately double the size of spleens from *B. vesperuginis*-free bats. Interestingly, there was no significant difference in spleen size between infected and uninfected juvenile bats, which suggests that splenomegaly is likely to be associated with repeat *B. vesperuginis* infections (Lord 2010).

4.2.3 Haemosporidia *in UK Bats*

In Europe, bats are known to be infected with the haemosporidians *Polychromophilus melanipherus* and *P. murinus* (Dionisi 1899). Although related to malaria parasites, these bat haemosporidia appear to have no detrimental health impact upon their hosts. In the UK, surveys show that only *P. murinus* is present; moreover, the Gardner and Molyneux study demonstrates that this parasite is highly host restricted since it was only documented in blood smears taken from *Myotis daubentonii* (Table 4.2) (Gardner 1986; Gardner and Molyneux 1988a). Moreover, attempts to inoculate captive pipistrelle bats with *P. murinus* sporozoites extracted from the salivary glands of the vector, the wingless bat fly *Nycteribia kolenatii*, proved unsuccessful (Gardner and Molyneux 1988a). Not surprisingly, regional differences were recorded based upon sampling location and season; for example, 34 % (10/29) infection in a Northamptonshire summer roost and 20 % (12/60) in hibernating bats sampled at sites across Norfolk and Suffolk (Gardner 1986). However, as no colony was sampled throughout the year, it was not possible to conclude anything about variations in parasite transmission between the seasons (Gardner and Molyneux 1988a).

Molecular-based studies to characterise *Polychromophilus* spp. infections have recently been carried out using PCR primers to amplify a fragment of the highly conserved apicomplexan mitochondrial cytochrome *b* gene from bats in Asia, Africa (Duval et al. 2007, 2012) and Europe (Megali et al. 2011). These studies have confirmed that *Polychromophilus* spp. is globally distributed and also that *P. murinus* is most closely associated with *M. daubentonii* (51 %, 65/127). Interestingly, *P. murinus* infections were also noted in *E. serotinus* (11 %, 2/18), *N. noctula* (7 %, 1/15) and *M. myotis* (4 %, 2/47); the authors concluding that the vector must be a nycteribiid other than *N. kolenatii* due to the highly specific associations that occur between these ectoparasites and their bat hosts (Megali et al. 2011). Gardner and Molyneux analysed 27 *N. noctula* and 15 *E. serotinus* blood smears and did not detect *P. murinus* (Gardner 1986). They concluded that because *N. kolenatii* is the only common nycteribiid in the UK, there is limited, if any, possibility of transmission to other UK bat species since nycteribiids are highly host-specific ectoparasites (Gardner and Molyneux 1988a).

Comparison of the *P. murinus* cytochrome *b* gene sequences from parasites derived from bats sampled in Switzerland demonstrated that they could vary by up to five nucleotides between isolates (Megali et al. 2011). Interestingly, different parasite haplotypes associated with *M. daubentonii* were present within the same colony. Moreover, the identical parasite haplotypes were also noted for individual bats from differing colonies. These gene sequences, along with more recent genetic data from *Polychromophilus* spp. (Witsenburg et al. 2012), are providing valuable contributions to our understanding of the evolution of the haemosporidia including the human malaria parasite *Plasmodium falciparum*. Indeed, the phylogenetic data supports a host switching event having occurred between an avian, or reptilian, ancestor of *Polychromophilus* spp. into bats with rapid adaptation to a nycteribiid vector (Witsenburg et al. 2012). As a consequence of this host switching, *Polychromophilus* spp. remains restricted to bats due to the highly specific association of the nycteribiid with its host. Since this bat–nycteribiid–parasite relationship is so highly specialised, as stated by Megali and colleagues, it is undoubtedly worthy of further study (Megali et al. 2011).

4.3 Gastrointestinal Parasites

The published literature shows that bats are infected with a plethora of gastrointestinal parasites; indeed, many summary records of bat nematode, cestode and trematode infections are readily accessible via the Natural History Museum Parasite Database (Gibson et al. 2005a). The research literature on bat helminths is dominated by records of species, new host and locality records and descriptions of new parasite species (Botella et al. 1993; Esteban et al. 1990; Lotz and Font 1991; Marshall and Miller 1979; Nahhas et al. 2005; Ricci 1995; Shimalov et al. 2002; Webster and Casey 1973). Despite these efforts, there is a paucity of knowledge about bat helminth community composition and the factors that influence parasite community structure. Furthermore, life cycle details, including basic information such as the species identities of intermediate hosts, are unknown for many of the parasites. To our knowledge, the majority of gastrointestinal parasite infections have little pathological consequences for their bat host (see Sect. 4.5).

Eimeria is the most speciose genus of coccidian parasite. Despite the estimated large numbers of eimerian species, very little work has been carried out on these parasites in bats. Duszynski (2002) has provided a comprehensive review of bat eimerian research and advocates that concerted, multidisciplinary efforts are needed to more fully understand the Eimeriidae and the relationship between bats and coccidia. Invasive techniques would be needed to study intracellular parasite development; however, non-invasive sampling, from guano, would allow an infection to be documented and potentially, questions about oocyst development to be addressed. As in other mammals, heavy eimerian infections are likely to cause damage to the intestinal epithelial cells of the host. Due to lack of bat eimerian studies though the precise extent of any health impact is largely unknown (see Sect. 4.5). Although most coccidia are localised to the gastrointestinal tract, some are known to cause renal coccidiosis; the causative agent in bats being postulated as *Eimeria* spp. based upon an unusual cystic dilation of the kidney tubules (Gruber et al. 1996). The health impact of renal coccidiosis upon the bat host remains unclear though it seems highly plausible that infection will elevate the risk of mortality. Indeed, a study of disease in almost 500 wild bats diagnosed renal coccidiosis in 11 specimens and approximately half of these bats were noted as being in 'poor' body condition (Mühldorfer et al. 2011a).

4.3.1 Helminths in UK Bats

Database searches for records of helminth parasites characterised from UK bats indicated that there were no public documents available, though helminths had been noted in bats during autopsy (personal communication, Vic Simpson, Wildlife Veterinary Investigation Centre, Truro, Cornwall). We therefore examined a collection of 99 pipistrelle bats sampled from across South Lancashire and Greater Manchester in order to generate some baseline data on gastrointestinal helminths in these hosts (Lord et al. 2012). Our findings showed that UK pipistrelles are commonly infected with trematodes (76 % prevalence) and that the parasites were distributed in an overdispersed pattern across the population (range 0–328 helminths per host; average abundance 48.2 $\pm$ 7). From the published bat parasite literature, trematodes appear to be the dominant helminth associated with bats; indeed, our analysis did not reveal any nematode, cestode or acanthocephalan infections.

We speciated the trematodes in a subset of pipistrelles, using a combination of morphological analysis and DNA sequencing of the 28S rRNA gene of representative specimens (Lord et al. 2012). The data showed that lecithodendriids are the most common helminths in pipistrelle bats and we also confirmed a relatively high prevalence of *Plagiorchis koreanus* (Table 4.3) (Figs. 4.1, 4.2, 4.3, 4.4, and 4.5). *Lecithodendrium linstowi*, significantly the most prevalent and abundant parasite, was always localised to the duodenum and upper sections of the jejunum. The remaining four species, *L. spathulatum*, *Pycnoporus heteroporus*, *Prosthodendrium* sp. and *P. koreanus*, were always localised to the ileum.

Although one pipistrelle was infected with all of the lecithodendriids and also *P. koreanus*, most of the bats harboured two species. The most common association was between *L. linstowi* and *L. spathulatum*. Other positive associations were noted between L. linstowi—*P. heteroporus*, L. linstowi—*Prosthodendrium* sp. and L. spathulatum—*P. koreanus* (Lord et al. 2012). Such associations are likely to be a consequence of these parasites sharing common intermediate hosts; in particular, aquatic insect larvae, which upon development to adulthood, become a potential food source for the foraging bat.

Statistical modelling of the parasite data showed that although there was no difference in overall prevalence between the sexes, intriguingly, male bat infections were significantly more aggregated than the distribution observed in females and also, significantly less abundant compared to females (Lord et al. 2012). This is indicative of subtle differences in infection occurring between the sexes, which may be explained by differential helminth responses to the male and female sex hormones (Haukisalmi et al. 1988). However, for most reported investigations, helminth infections of vertebrates are usually more severe in males than females, with males often harbouring higher worm burdens (Poulin 1996; Wedekind and Jacobsen 1998). Studies on helminth infections in other mammals (Klein 2004) may inform hypothesis testing approaches for understanding the sex-biased data that we have identified in bats (Lord et al. 2012). However, this will undoubtedly

Table 4.3 Trematode infection data for a subset of 51 *P. pipistrellus* specimens collected across South Lancashire and Greater Manchester

Trematode species	Prevalence (%)	Mean abundance (±S.E.M.)
Lecithodendrium linstowi	80.4 (66.8–89.6)	50.7 ± 9.2
L. spathulatum	19.6 (10.4–33.1)	0.9 ± 0.3
Prosthodendrium sp.	35.3 (23.4–49.3)	2.3 ± 0.7
Plagiorchis koreanus	29.4 (18.4–43.1)	1.1 ± 0.5
Pycnoporus heteroporus	9.8 (4–21.3)	0.5 ± 0.3

Ninety five percent of confidence limits are given in parenthesis for the prevalence data. Table modified from Lord et al. (2012)

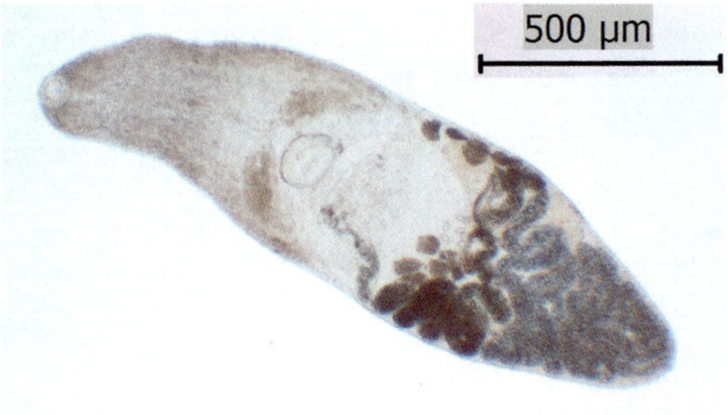

Fig. 4.1 *Pycnoporus heteroporus* (Image reproduced from Lord 2010)

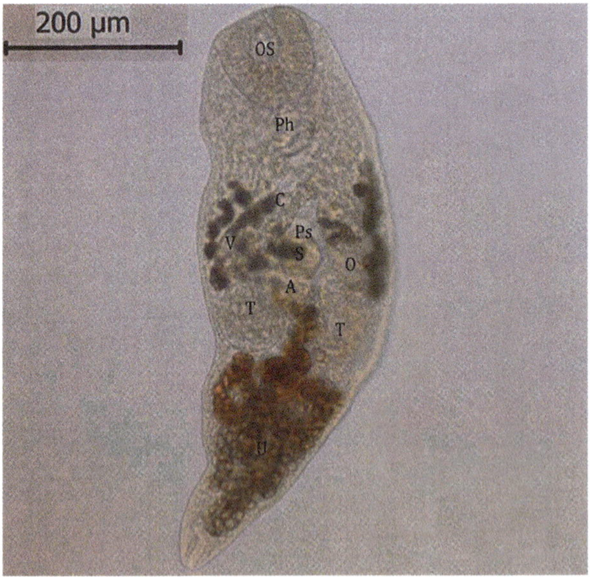

Fig. 4.2 *Prosthodendrium* sp. *Os* oral sucker, *Ph* pharynx, *A* acetabulum, *Ps* pseudobursa, *S* seminal vesicle, *T* testes, *O* ovary, *V* vitellaria, *U* uterus (Image reproduced from Lord 2010)

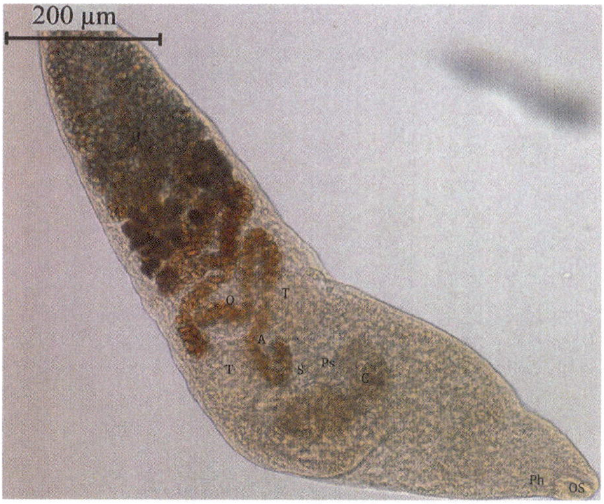

Fig. 4.3 *Lecithodendrium spathulatum*. *Os* oral sucker, *Ph* pharynx, *A* acetabulum, *Ps* pseudobursa, *C* caeca, *T* testes, *O* ovary, *U* uterus (Image reproduced from Lord 2010)

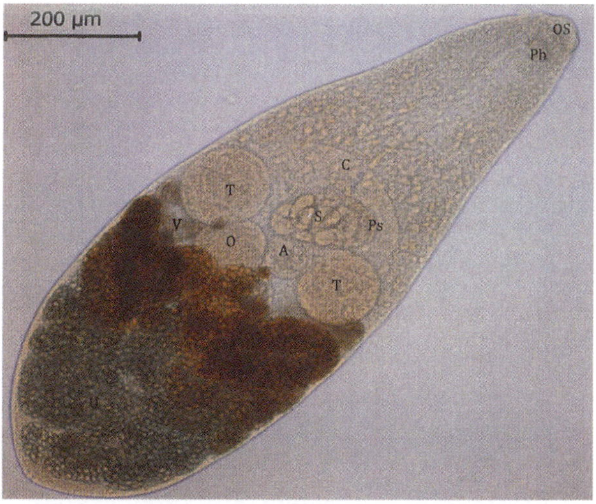

Fig. 4.4 *Lecithodendrium linstowi*. *Os* oral sucker, *Ph* pharynx, *A* acetabulum, *Ps* pseudobursa, *S* seminal vesicle, *C* caeca, *T* testes, *O* ovary, *V* vitellaria, *U* uterus (Image reproduced from Lord 2010)

present complex challenges given the protected species designation that governs bats.

Statistical modelling of the infection data also demonstrated that there was a significant increase in helminth prevalence and abundance throughout the bat sampling period (September 2005–September 2009) (Lord et al. 2012). Reasons for this are likely to be multifactorial, including changes to bat and intermediate host density, as well as climatic influences (e.g. rainfall). A possible seasonal infection trend, with helminth prevalence and abundance high in late spring, reducing throughout summer and increasing in early autumn, was not confirmed with statistical models due to the limited sample size. Nonetheless, the increased helminth prevalence and abundance in early autumn is consistent with summer

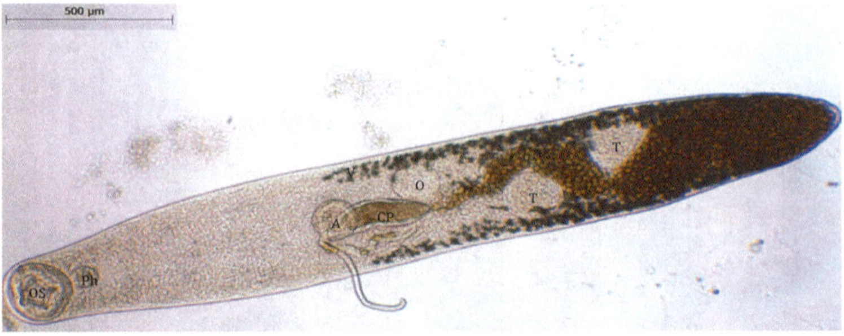

Fig. 4.5 *Plagiorchis koreanus*. Abbreviations as follows: *Os* oral sucker, *Ph* pharynx, *A* acetabulum, *CP* cirrus pouch, *T* testes, *O* ovary, *V* vitellaria (Image reproduced from Lord 2010)

being the peak period for transmission due to high numbers of intermediate insect hosts and a peak density of actively feeding bats. The late spring data may indicate that parasites are retained by the pipistrelles during hibernation.

As bat species are numerous and they occupy a unique niche for mammalian species, analysis of bat parasite DNA sequences can have profound influence on interpretation of evolutionary relationships as discussed for the bat haematozoa (Sects. 4.2.1 and 4.2.3). To this end, we reconstructed a phylogram of the bat Lecithodendriidae, incorporating novel 28S rRNA gene sequences obtained from representative samples of *L. spathulatum* and *Prosthodendrium* sp. The resulting data sheds new insight on relationships between the bat lecithodendriids since previous phylogenetic analyses had proposed the existence of a somewhat controversial clade between *L. linstowi* and *Prosthodendrium hurkovaae* (Tkach et al. 2003). Our analysis, using novel sequences, disrupts this clade to form separate *L. linstowi—L. spathulatum* and *Prosthodendrium* sp.—*P. hurkovaae* clades (Lord et al. 2012). Further sampling of bat specimens is however warranted to produce a more robust phylogram of the Lecithodendriidae.

Interestingly, *Neorickettsia risticii*, the gram-negative obligate intracellular bacterium responsible for Potomac Horse Fever, is known to associate with lecithodendriids (Gibson et al. 2005b; Pusterla et al. 2003). As many studies, including our own, have shown that these trematodes are highly prevalent and abundant in bats, it would be worthwhile investigating the role, if any, played by bats in the epidemiology of Potomac Horse Fever, a disease currently not documented in the UK but widespread in the USA. This could be initiated via diagnostic PCR screening of lecithodendriids isolated from bats for the presence of *N. risticii*.

4.3.2 Coccidia in UK Bats

To our knowledge, there are no published articles on eimerian infections in UK bats. To address this paucity of data, we utilised a combination of microscopic

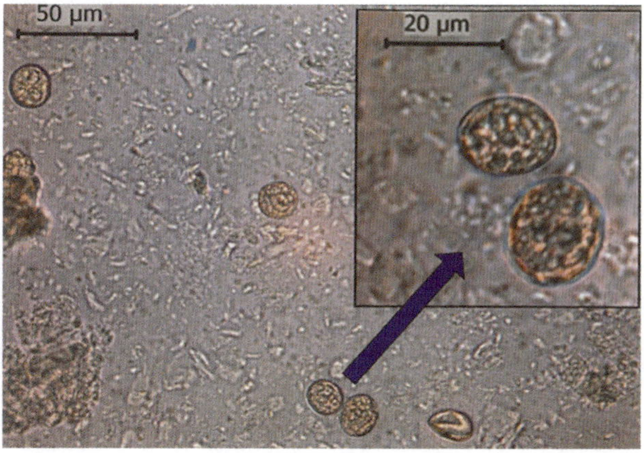

Fig. 4.6 Eimerian oocysts from bat faecal matter observed at ×400 and ×1,000 (oil immersion) magnification. Image reproduced from Lord (2010)

analysis of pipistrelle gut sections and faeces and also PCR screening of DNA extracted from gut tissue to diagnose eimerian parasite infections. Our data confirm the presence of *Eimeria* sp. in UK bats (20 % prevalence) (Fig. 4.6). Moreover, DNA sequence analysis of PCR amplified 18S rRNA gene fragments showed that the coccidian was highly related to *E. rioarribaensis*, an eimerian of *Myotis* spp. (Zhao et al. 2001). Further analyses are now warranted to more fully explore eimerian infections in UK bats.

4.4 *Toxoplasma gondii*: A Significant Natural Reservoir in Bats?

T. gondii is an obligate intracellular parasite of many species of warm blooded animals. Despite this, it has only recently been recognised that bats may be infected with *T. gondii* (Cabral et al. 2013; Yuan et al. 2013) and there is no published information on UK bat populations. As such, we examined the pipistrelle population sampled from South Lancashire and Greater Manchester for *T. gondii* infection using a PCR approach for detection of the parasite *SAG1* gene in DNA prepared from bat brain tissue. The data confirms that bats are not only infected with *T. gondii* but that levels of infection are surprisingly high (10 %) (Dodd et al. in preparation). This is likely to reflect that pipistrelles have successfully adapted to roosting in close proximity to humans and their domestic animals; not least cats, which are the definitive host of *T. gondii*. Bats may therefore have a significant role as natural reservoir hosts of *T. gondii*. Further studies are clearly warranted, not least to establish the mode of infection since pipistrelles are insectivorous and hence transmission can only be possible via oocyst ingestion and a congenital route. With regard to the former, one possibility is that bats may drink water infected with *T. gondii* oocysts. Alternatively, it is possible that aquatic insect larvae may be involved since any oocysts that they contain would be transmitted to the bat when foraging.

4.5 Bat Health and Parasite Infection

Although the focus of this chapter has been on UK bat parasites, it is not possible to overlook a recent significant research contribution that explored the reasons for death in almost 500 bats, representing 19 European vespertilionid species, sampled from across Germany (Mühldorfer et al. 2011a, b). Overall, 29 % of bats in the study had protozoan, or helminth, infections detectable by microscopic examination of organ tissues and statistical modelling of the data showed (1) a significant increase in parasite prevalence in older bats, (2) slightly increased infection prevalence in female bats compared to males and (3) increased prevalence in larger bat species compared to smaller species (Mühldorfer et al. 2011b).

With regard to causes of mortality, approximately one-third of the bat deaths were due to disease, and another one-third due to trauma; parasite infection, designated as "*severe intestinal trematode infection, disseminated nematode infection, renal or intestinal coccidiosis*," was observed in approximately 10 % of these cases (29/289) (Mühldorfer et al. 2011b). Moreover, parasitic infection was recorded as the primary cause of death for only two bats (~0.5 %) in the sampled population (Mühldorfer et al. 2011b). The parasites involved in these two cases were not identified to species level; however, it was evident that nematode larval migration was responsible for granulomas in multiple organs, including the brain, in a *M. myotis* specimen, whereas a *M. mystacinus* bat had severe intestinal coccidiosis (personal communication, Kristin Mühldorfer, Leibniz Institute for Zoo and Wildlife Research, Berlin). With respect to the disseminated nematode infection, it is distinctly possible that the bat may have been acting as a paratenic host.

Importantly, statistical modelling of the data confirmed that there was no significant association between the cause of bat death and the severity of parasite infection (Mühldorfer et al. 2011b). As such, this study provides compelling evidence that most endoparasite infections have limited impact upon bat health at the individual host level and hence negligible impact at the population level. For diseased bats, the greatest mortality risk is due to bacterial infection (12.5 % mortality), while viral infection (AdV-2 and EBLV-1) presents an almost 2.5 times greater risk of mortality compared to parasite infection (Mühldorfer et al. 2011b).

4.6 Conclusions

Due to their life histories, which facilitate high parasite transmission rates, bats are undoubtedly a rich source of parasite biodiversity and this itself, most likely reflects a healthy ecosystem (Hudson et al. 2006). There are many areas of biology that would benefit from more detailed studies of bat parasites; however, this will undoubtedly be challenging given that protected species designation makes invasive sampling of bats difficult. As such, further studies are likely to focus on alternative approaches. For example, the examination of bats reported dead, or euthanized due to injury, or ill health, by a qualified person. Unfortunately, the quality of parasite material recovered by this approach is frequently less than optimal and hence speciation of the parasites, many of which are rarely encountered and hence not well described, is severely compromised. Alternatively, non-invasive sampling strategies should allow collection and examination of bat parasites. Non-invasive sampling should include the dissection and examination of bat ectoparasites and also the collection and analyses of guano for gastrointestinal parasite analyses.

These approaches require detailed planning and cooperation between bat group volunteers, veterinarians and scientists in order to minimise impact upon the host. Moreover, prior to commencement of any work on UK bats, including dead specimens, the appropriate licenses must be obtained. Given the potential infection risk due to bat lyssavirus, an appropriate risk assessment must also be completed. Rabies monitoring in the UK is carried out by the Veterinary Laboratory Agency and hence one approach is to delay bat autopsy and parasite investigation for a suitable period of time in order to allow lyssavirus testing of the specimen.

The recent molecular-based surveys of UK bat haemoparasites (Lord 2010; Concannon et al. 2005) and helminths (Lord et al. 2012) are useful demonstrations of how scientists, veterinarians and bat group workers can collaborate to further knowledge of bat parasites. This work has confirmed, as reported by studies elsewhere, that UK bats are host to a variety of helminth and protozoan parasites. Indeed, *T. gondii* has only recently been recognised as being associated with bats (Dodd et al. in preparation) and further extensive investigations are likely to describe new strains of other parasites [e.g. trypanosomes (Hamilton et al. 2012b)]. Continued collaborative partnerships will be necessary to pursue future goals, both in the UK and elsewhere.

4.6.1 Future Perspectives

Given that bats are a rich source of parasites, many of which appear to be highly host specific, then more rigorous and complete species descriptions are necessary from widely sampled hosts. Even when parasite species are reasonably well described, there is often a lack of reference material and an absence of DNA sequence data that precludes more detailed investigations (e.g. *Babesia vesperuginis*). Development of in vitro parasite cultivation systems would contribute enormously to more thorough investigations; however, this will undoubtedly remain a major challenge for many species. As such, a more realistic aim would be to continue to characterise more readily accessible parasites, particularly helminths given that DNA can be isolated from individual specimens and used to assist identification and phylogenetic studies. Cataloguing gastrointestinal helminth species in this manner will also allow a more rigorous examination of parasite community composition and the factors that may influence it.

The identities of intermediate host species for many bat parasites remain unknown. To this end, field sampling of invertebrates for parasite examination is a necessity. For example, the trematode species described in our study of pipistrelle bats (Lord et al. 2012) would be predicted to develop via an aquatic snail and then an aquatic insect intermediate host. The utilisation of parasite PCR screening approaches on field collected aquatic snails and insects should efficiently identify potential intermediate host candidates. Further field material could then be collected and studied more extensively under laboratory conditions to robustly establish the relationship between parasite and intermediate hosts.

Bats have been the focus of much attention in recent years due to their association with emerging zoonotic diseases of viral origin (Calisher et al. 2006; Luis et al. 2013). It should be stressed that there is currently no zoonotic concern with respect to bat endoparasites. It should also be noted that bat endoparasites appear to have very limited health impacts upon their host (Mühldorfer et al. 2011a, b), which is usually indicative of well-established evolutionary relationships. Nonetheless, increasing environmental stressors, including habitat loss and habitat change, may have unpredictable impacts upon bats, their parasites and other infectious agents (viral, bacterial and fungal) that they harbour (Acevedo-Whitehouse and Duffus 2009). Improved understanding of how these microbial communities interact with each other and their host is fundamental to predicting potential outcomes, both for the host and the wider ecosystem (Hayman et al. 2013). To this end, continual monitoring and study of bat endoparasites, alongside a greater understanding of host biology, including genetic and immunological studies (Baker et al. 2013), in addition to ecology-based research, is of paramount importance. This will clearly require multidisciplinary efforts if bat parasitology is to continue to make useful and significant contributions to our understanding of wildlife health, disease and biodiversity.

Acknowledgements We thank the University of Salford for funding JSL's PhD studentship and members of the South Lancashire Bat Group for assistance with acquisition of pipistrelles for parasite study in our laboratory. The focus of this chapter has been on UK bat parasites; however, we gratefully acknowledge the significant contributions made to bat parasitology research by the wider community.

References

Acevedo-Whitehouse K, Duffus AL (2009) Effects of environmental change on wildlife health. Philos Trans R Soc Lond B Boil Sci 364(1534):3429–3438

Anez N, Crisante G, Soriano PJ (2009) *Trypanosoma cruzi* congenital transmission in wild bats. Acta Trop 109:78–80

Baker JR (1973) First european record of *Trypanosoma (Megatrypanum)* sp. of bats. Nat New Biol 241(107):96

Baker JR (1974) Protozoan parasites of the blood of British wild birds and mammals. J Zool 172:169–190

Baker JR, Thompson GB (1971) Two species of *Trypanosoma* from British bats. Trans R Soc Trop Med Hyg 65(4):427

Baker JR, Green SM, Chaloner LA, Gaborak M (1972) *Trypanosoma (Schizotrypanum) dionisii* of *Pipistrellus pipistrellus* (Chiroptera): intra- and extracellular development in vitro. Parasitology 65:251–263

Baker ML, Schountz T, Wang LF (2013) Antiviral immune responses of bats: a review. Zoonoses Public Health 60:104–116

Barratt EM, Deaville R, Burland TM, Bruford MW, Jones G, Racey PA, Wayne RK (1997) DNA answers the call of pipistrelle bat species. Nature 387:138–139

Bat Conservation Trust (2011) The state of the UK's bats. National bat monitoring programme population trends 2011. Joint Nature Conservation Committee, Peterborough

Botella P, Sanchez L, Esteban J (1993) Helminthfauna of bats in Spain. III. Parasites of *Pipistrellus pipistrellus* (Schreber, 1774) (Chiroptera: Vespertilionidae). Res Rev Parasitol 53:63–70

Cabral AD, Gama AR, Sodre MM, Savani ES, Galvao-Dias MA, Jordao LR, Maeda MM, Yai LE, Gennari SM, Pena HF (2013) First isolation and genotyping of *Toxoplasma gondii* from bats (Mammalia: Chiroptera). Vet Parasitol 193:100–104

Calisher CH, Childs JE, Field HE, Holmes KV, Schountz T (2006) Bats: important reservoir hosts of emerging viruses. Clin Microbiol Rev 19:531–545

Coles AC (1914) Blood parasites found in mammals, birds, and fishes in England. Parasitology 7:17–61

Concannon R, Wynn-Owen K, Simpson VR, Birtles RJ (2005) Molecular characterization of haemoparasites infecting bats (Microchiroptera) in Cornwall, UK. Parasitology 131:489–496

Dionisi A (1898) Les parasites endoglobulaires des chauves-souris. Atti della Reale Academia dei Lincei 7:153–156

Dionisi A (1899) La malaria di alcuni specie di pipistrelli. Atti della Societa per gli studi Malaria 1:133–173

Dodd NS, Lord JS, Jehle R, Brooks,DR, Hide G (in preparation) Prevalence of *Toxoplasma gondii* in two populations of British bats (*Pipistrellus pipistrellus and P. pygmaeus*)

Duszynski DW (2002) Coccidia (Apicomplexa: Eimeriidae) of the mammalian order Chiroptera. Mus Southwest Biol 5:1–45

Duval L, Robert V, Csorba G, Hassanin A, Randrianarivelojosia M, Walston J, Nhim T, Goodman SM, Ariey F (2007) Multiple host-switching of Haemosporidia parasites in bats. Malar J 6:157

Duval L, Mejean C, Maganga GD, Makanga BK, Mangama Koumba LB, Peirce MA, Ariey F, Bourgarel M (2012) The chiropteran haemosporidian *Polychromophilus melanipherus*: a worldwide species complex restricted to the family Miniopteridae. Infect Genet Evol 12:1558–1566

Esteban JG, Oltra Ferrero JL, Mas-coma S (1990) Helmintofauna de los murcielagos de Espana. II. Parasitos de *Miniopterus schreibersi* (Kuhl, 1819) (Chiroptera: Vespertilionidae). Rev Iber Parasitol 50:199–209

Fooks AR, McElhinney LM, Pounder DJ, Finnegan CJ, Mansfield K, Johnson N, Brookes SM, Parsons G, White K, McIntyre PG, Nathwani D (2003) Case report: isolation of a European bat lyssavirus type 2a from a fatal human case of rabies encephalitis. J Med Virol 71:281–289

Frick WF, Pollock JF, Hicks AC, Langwig KE, Reynolds DS, Turner GG, Butchkoski CM, Kunz TH (2010) An emerging disease causes regional population collapse of a common North American bat species. Science 329:679–682

Garcia HA, Rodrigues AC, Martinkovic F, Minervino AH, Campaner M, Nunes VL, Paiva F, Hamilton PB, Teixeira MM (2011) Multilocus phylogeographical analysis of *Trypanosoma (Megatrypanum)* genotypes from sympatric cattle and water buffalo populations supports evolutionary host constraint and close phylogenetic relationships with genotypes found in other ruminants. Int J Parasitol 41:1385–1396

Gardner R (1986) Studies on the haematozoa of British bats. Ph.D. thesis, University of Salford

Gardner RA, Molyneux DH (1987) *Babesia vesperuginis*: natural and experimental infections in British bats (Microchiroptera). Parasitology 95:461–469

Gardner RA, Molyneux DH (1988a) *Polychromophilus murinus*: a malarial parasite of bats: life-history and ultrastructural studies. Parasitology 96:591–605

Gardner RA, Molyneux DH (1988b) *Schizotrypanum* in British bats. Parasitology 97:43–50

Gardner RA, Molyneux DH (1988c) *Trypanosoma (Megatrypanum) incertum* from *Pipistrellus pipistrellus*: development and transmission by cimicid bugs. Parasitology 96:433–447

Gardner RA, Molyneux DH, Stebbings RE (1987) Studies on the prevalence of haematozoa of British bats. Mammal Rev 17:75–80

Gibson DI, Bray RA, Harris EA (2005a) Host-parasite database of the Natural History Museum, London

Gibson KE, Rikihisa Y, Zhang C, Martin C (2005b) *Neorickettsia risticii* is vertically transmitted in the trematode *Acanthatrium oregonense* and horizontally transmitted to bats. Environ Microbiol 7:203–212

Gruber AD, Schulze CA, Brugmann M, Pohlenz J (1996) Renal coccidiosis with cystic tubular dilatation in four bats. Vet Pathol 33:442–445

Hale JD, Fairbrass AJ, Matthews TJ, Sadler JP (2012) Habitat composition and connectivity predicts bat presence and activity at foraging sites in a large UK conurbation. PLoS One 7: e33300

Hamilton PB, Teixeira MM, Stevens JR (2012a) The evolution of *Trypanosoma cruzi*: the 'bat seeding' hypothesis. Trends Parasitol 28:136–141

Hamilton PB, Cruickshank C, Stevens JR, Teixeira MM, Mathews F (2012b) Parasites reveal movement of bats between the New and Old Worlds. Mol Phylogenet Evol 63:521–526

Harris S, Morris P, Wray S, Yaldon D (1995) A review of British mammals: population estimates and conservation status of British mammals other than cetaceans. Joint Nature Conservation Committee, Peterborough

Harris SL, Aegerter JN, Brookes SM, McElhinney LM, Jones G, Smith GC, Fooks AR (2009) Targeted surveillance for European bat lyssaviruses in English bats (2003–06). J Wildl Dis 45:1030–1041

Haukisalmi V, Henttonen H, Tenora F (1988) Population dynamics of common and rare helminths in cyclic vole populations. J Anim Ecol 57:807–825

Hayman DT, Bowen RA, Cryan PM, McCracken GF, O'Shea TJ, Peel AJ, Gilbert A, Webb CT, Wood JL (2013) Ecology of zoonotic infectious diseases in bats: current knowledge and future directions. Zoonoses Public Health 60:2–21

Hudson PJ, Dobson AP, Lafferty KD (2006) Is a healthy ecosystem one that is rich in parasites? Trends Ecol Evol 21:381–385

Hutson AM (1984) Keds, flat-flies and bat-flies (Hippoboscidae and Nycteribiidae), vol 10. Royal Entomological Society, London

Hutson AM, Aulagnier S, Spitzenberger F (2008a) *Barbastella barbastellus*. IUCN 2012 IUCN Red List of threatened species

Hutson AM, Spitzenberger F, Tsytsulina K, Aulagnier S, Juste J, Karataş A, Palmeirim J, Paunović M (2008b) *Myotis bechsteinii*. IUCN 2012 IUCN Red List of threatened species

Klein SL (2004) Hormonal and immunological mechanisms mediating sex differences in parasite infection. Parasite Immunol 26:247–264

Lima L, Silva FM, Neves L, Attias M, Takata CS, Campaner M, de Souza W, Hamilton PB, Teixeira MM (2012) Evolutionary insights from bat trypanosomes: morphological, developmental and phylogenetic evidence of a new species, *Trypanosoma (Schizotrypanum) erneyi* sp. nov., in African bats closely related to *Trypanosoma (Schizotrypanum) cruzi* and allied species. Protist 163:856–872

Lord JS (2010) Micro and macroparasites of bats (Chiroptera). Ph.D. thesis, University of Salford

Lord JS, Parker S, Parker F, Brooks DR (2012) Gastrointestinal helminths of pipistrelle bats (*Pipistrellus pipistrellus*/*Pipistrellus pygmaeus*) (Chiroptera: Vespertilionidae) of England. Parasitology 139:366–374

Lotz JM, Font WF (1991) The role of positive and negative interspecific associations in the organization of communities of intestinal helminths of bats. Parasitology 103:127–138

Luis AD, Hayman DT, O'Shea TJ, Cryan PM, Gilbert AT, Pulliam JR, Mills JN, Timonin ME, Willis CK, Cunningham AA, Fook AR, Rupprecht CE, Wood JLN, Webb CT (2013) A comparison of bats and rodents as reservoirs of zoonotic viruses: are bats special? Proc R Soc B 280:20122753

Marshall M, Miller G (1979) Some digenetic trematodes from Ecuadorian bats including five new species and one new genus. J Parasitol 65:909–917

Megali A, Yannic G, Christe P (2011) Disease in the dark: molecular characterization of *Polychromophilus murinus* in temperate zone bats revealed a worldwide distribution of this malaria-like disease. Mol Ecol 20:1039–1048

Molyneux DH (1992) Trypanosomes of bats. In: Kreier JP, Baker JR (eds) Parasitic protozoa. Academic, New York, NY, pp 195–223

Molyneux DH, Baffort JM (1971) Observations on the trypanosome of *Pipistrellus pipistrellus* in Britain, *Trypanosoma (Schizotrypanum) vespertilionis*. Ann Soc Belg Med Trop 51:335–348

Mühldorfer K, Speck S, Wibbelt G (2011a) Diseases in free-ranging bats from Germany. BMC Vet Res 7:61

Mühldorfer K, Speck S, Kurth A, Lesnik R, Freuling C, Muller T, Kramer-Schadt S, Wibbelt G (2011b) Diseases and causes of death in European bats: dynamics in disease susceptibility and infection rates. PLoS One 6:e29773

Nahhas FA, Yang P, Uch S (2005) Digenetic trematodes of *Tadarida brasiliensis mexicana* (Chiroptera: Molossidae) and *Myotis californicus* (Chiroptera: Vespertilionidae) from Northern California, U.S.A. Comp Parasitol 72:196–199

Petrie GF (1905) Observations relating to the structure and geographical distribution of certain Trypanosomes. J Hyg 5:191–200

Poulin R (1996) Helminth growth in vertebrate hosts: does host sex matter? Int J Parasitol 26:1311–1315

Pusterla N, Johnson EM, Chae JS, Madigan JE (2003) Digenetic trematodes, *Acanthatrium* sp. and *Lecithodendrium* sp., as vectors of *Neorickettsia risticii*, the agent of Potomac horse fever. J Helminthol 77:335–339

Ramsey JM, Gutierrez-Cabrera AE, Salgado-Ramirez L, Peterson AT, Sanchez-Cordero V, Ibarra-Cerdena CN (2012) Ecological connectivity of *Trypanosoma cruzi* reservoirs and *Triatoma pallidipennis* hosts in an anthropogenic landscape with endemic Chagas disease. PLoS One 7: e46013

Ricci M (1995) Trematode parasites of Italian bats. Parassitologia 37:199–214

Rivers NM, Butlin RK, Altringham JD (2005) Genetic population structure of Natterer's bats explained by mating at swarming sites and philopatry. Mol Ecol 14:4299–4312

Shimalov VV, Demyanchik MG, Demyanchik VT (2002) A study on the helminth fauna of the bats (Mammalia, Chiroptera: Vespertilionidae) in Belarus. Parasitol Res 88:1011

Simpson VR (2000) Veterinary advances in the investigation of wildlife diseases in Britain. Res Vet Sci 69:11–16

Tkach VV, Littlewood DT, Olson PD, Kinsella JM, Swiderski Z (2003) Molecular phylogenetic analysis of the Microphalloidea Ward, 1901 (Trematoda: Digenea). Syst Parasitol 56:1–15

Wang LF, Walker PJ, Poon LL (2011) Mass extinctions, biodiversity and mitochondrial function: are bats 'special' as reservoir for emerging viruses? Curr Opin Virol 1:649–657

Webster WA, Casey GA (1973) Studies on the parasites of Chiroptera. III. Helminths from various bat species collected in British Columbia. Can J Zool 51:633–636

Wedekind C, Jacobsen PJ (1998) Male-biased susceptibility to helminth infection: an experimental test with a copepod. Oikos 81:458–462

Witsenburg F, Salamin N, Christe P (2012) The evolutionary host switches of *Polychromophilus*: a multi-gene phylogeny of the bat malaria genus suggests a second invasion of mammals by a haemosporidian parasite. Malar J 11:53

Yuan ZG, Luo SJ, Dubey JP, Zhou DH, Zhu YP, He Y, He XH, Zhang XX, Zhu XQ (2013) Serological evidence of *Toxoplasma gondii* infection in five species of bats in China. Vector Borne Zoonotic Dis 13(6):422–424. doi:10.1089/vbz.2012.1091

Zhao X, Duszynski DW, Loker ES (2001) Phylogenetic position of *Eimeria antrozoi*, a bat coccidium (Apicomplexa: Eimeriidae) and its relationship to morphologically similar *Eimeria* spp. from bats and rodents based on nuclear 18S and plastid 23S rDNA sequences. J Parasitol 87:1120–1123

Chapter 5
Macroparasites of Microchiroptera: Bat Ectoparasites of Central and South America

Raphael Frank, Julian Münster, Julia Schulze, Andrew Liston, and Sven Klimpel

Abstract The highest density of bat species in the world is reached in South and Central America. Representatives of nine families occur in this gigantic region between the southern part of the continental twin and the land bridge to North America. They display a high degree of ecological variance: exemplified by species with extensive or endemic distributions, specialist and generalist feeding habits, colonial and solitary roosting, and those that are migratory or permanently resident. However, they are all exposed to the deleterious effects of parasites. In terms of the number of species and higher taxa, the diversity of parasites considerably exceeds that of their hosts. A total of 172 bat species of 9 families as hosts and 273 ectoparasite species are summarised in this review. We included data from 20 South and Central American countries. Of foremost importance are Diptera of the families Nycteribiidae and Streblidae with 187 different species. For the last mentioned family, we recorded more than 61 % of the worldwide known species. Further, fleas (Siphonaptera) (8 species), bugs (Hemiptera) (4 species) and Arachnida of the orders Ixodida and Mesostigmata (74 species) are also represented. The large numbers of species, like *Basilia carteri*, belong to the two families of Diptera, which parasitise only on bats. Together with their high degree of specialisation, it is an indication for their high rate of adaptation. Investigations on the parasite fauna of bats have been conducted in the named areas with different frequency and intensity. This chapter contains a reference list of the bats that have been scientifically studied in South and Central America so far, with respect to their associated parasites. We

R. Frank (✉) • J. Münster • J. Schulze • S. Klimpel (✉)
Biodiversity and Climate Research Centre (BiK-F), Medical Biodiversity and Parasitology; Senckenberg Gesellschaft für Naturforschung (SGN); Goethe-University (GO), Institute for Ecology, Evolution and Diversity, Senckenberganlage 25, 60325 Frankfurt am Main, Germany
e-mail: Raphael.Frank@senckenberg.de; klimpel@bio.uni-frankfurt.de

A. Liston
Senckenberg German Entomological Institute (SDEI), Eberswalder Str. 90, 15374 Muencheberg, Germany

S. Klimpel and H. Mehlhorn (eds.), *Bats (Chiroptera) as Vectors of Diseases and Parasites*, Parasitology Research Monographs 5, DOI 10.1007/978-3-642-39333-4_5,

also present the results of our own study on the ectoparasites of bats in Bolivia, where 16 parasite species from 10 different host species were identified.

Keywords Bat ectoparasites • Microchiroptera • Nycteribiidae • Streblidae

5.1 Introduction

Bats are the second largest group of mammals on earth. They are distributed worldwide but are most abundant in tropical habitats, and only absent in the polar regions. More than 1,000 species in 18 families are currently known (Simmons 2005). Species richness of bats reaches astonishing dimensions in South and Central America. For example, this diversity is reflected in an area of Guyana where significantly more than half of the mammal species are bats (Lim and Engstrom 2004). This dominance within the mammalian fauna is typical for lowland neotropical rainforests. Their marked ecological success is, amongst other factors, attributable to specialisation on particular sources of food, their social behaviour, as well as their capability of active flight (Patterson et al. 2008). These numerous bat species also serve as hosts for a multitude of parasites (Bertola et al. 2005). In recent years, the interest on works dealing with bats and their role as hosts to numerous parasitic organisms, including metazoan ecto- and endoparasites, has increased. Several studies have revealed that host specificity is high, even under the most unlikely conditions that occur in tropical bat communities (Dick 2007; Dick and Patterson 2007). Because the hosts are highly mobile, roost in large groups comprising different families in close body contact, and given that parasitic organisms such as bat flies are equally mobile, a low specialisation of the parasites could be expected. One explanation among others for the development of such a high number of monoxenous parasite species is based on the immunological adaptation. This adaption between hosts and their parasites leads to a reduced or circumvened immunological reaction, which therefore reduces the risk of physical defence by the host (Dick and Patterson 2007; Fritz 1983; Khokhlova et al. 2004; Mans et al. 2002; Salzet et al. 2000).

This hypothesis, that even the small numbers of bat fly species which are oligoxenous or polyxenous have their preferred host species, is supported by the findings of Dick (2007). Host switching occurs only to bats of the same genus (oligoxenous) or family (polyxenous). Taking these observations into account, it is interesting that Patterson et al. (1998) found distributional evidence for co-speciation between New World bats and their associated bat flies. The phylogeny of nycteribiid and streblid bat flies is correlated with that of their hosts, what is also shown by the clear taxonomic division into a New World and an Old World clade within the flies (Patterson et al. 1998; Dittmar et al. 2006). Matching divisions can be seen in the phylogeny of the bats. Of all 18 families of the order Chiroptera, eight occur only in the Old World, seven are restricted to the New World and only three families have a worldwide distribution (Eick et al. 2005; Teeling et al. 2005).

These observations refer of course to a large scale of geographical area. However, Bruyndonckx et al. (2009, 2010) found hints of co-speciation even between European bats and their ectoparasitic mites (Guiller and Deunff 2010). On this smaller scale, patterns are not always as clear as demonstrated by a study on co-speciation patterns of nycteribiid bat flies and their host bats in Japan. In this study, no correlations between the respective phylogenies were found (Nikoh et al. 2011). Despite the fact that there is an increasing interest in bat and parasite phylogeny and host–parasite interactions, there is still little known about the processes, time scale and degree of bat ectoparasite specialisation.

Species of the bat-specific dipterous families Nycteribiidae (monophyletic) and Streblidae (paraphyletic) are by far the most abundant and also the best adapted ectoparasites of bats (Dittmar et al. 2006). A few species of Streblidae display a tendency towards mesoparasitism (Linhares and Komeno 2000). The oldest fossil records of Streblidae can be dated to 15 million years ago (Poinar and Brown 2012). The taxonomic position of the sister families Nycteribiidae and Streblidae is controversial. Here, we followed the generally accepted classification, in which they are placed in the superfamily Hippoboscoidea (formerly known as Pupipara) (Yeates and Wiegmann 1999). One of the main reasons for this placement is the adenotrophic viviparity of both families. Regional as well as national reference works and various published studies deal with the ecological impact of these parasites on their hosts (Hofstede and Fenton 2005; Patterson et al. 1998; Wenzel 1976; Wenzel and Tipton 1966). At least 286 species of Nycteribiidae are known, which belong to the three subfamilies (Graciolli et al. 2007): Archynycteribiinae (1 genus, 3 species), Cyclopodiinae (4 genera, 60 species) and Nycteribiinae (6 genera, 212 species) (Autino et al. 2009; Dick and Miller 2010). The first two subfamilies are confined to the Old World and are predominantly parasites of Megachiroptera. On the other hand, representatives of the Nycteribiinae occur in the Old and New Worlds. Within the Streblidae, at least 229 species of Streblidae are known so far with a worldwide distribution (Dick and Miller 2010). Furthermore, five subfamilies are distinguished, each with a clearly delimited geographic range: Nycterophiliinae (2 genera, 6 species), Streblinae (4 genera, 35 species) and Trichobiinae (20 genera, 115 species) are restricted to the New World (Dick and Miller 2010), while the Brachytarsininae (4 genera, 52 species) and Ascodipterinae (3 genera, 21 species) occur only in the Old World.

As a result of the species richness and widespread distribution of the Nycteribiidae and Streblidae, they form a large part of the metazoan parasite fauna of bats (Dick 2006; Graciolli 2004; Hofstede et al. 2004; Wenzel and Tipton 1966; Wenzel 1976).

Fleas (order Siphonaptera) are more rarely recorded on bats (Autino et al. 2009, 2011; Munoz et al. 2011; Wenzel and Tipton 1966). Those which are parasitic on bats belong to the families Ischnopsyllidae (20 genera, 122 species), Stephanocircidae (9 genera, 51 species) and Tungidae (4 genera) (Autino et al. 2011; Whiting et al. 2008). The Ischnopsyllidae parasitise exclusively bats and are distributed worldwide. Stephanocircidae are distributed in Australia and the Nearctic. Their host range includes Metatheria and Rodentia. Tungidae parasitise not only diverse mammals, including bats and rodents, but also birds (Whiting et al. 2008) and occur in the Neotropics, Africa and Asia. Additionally, true bugs (order Hemiptera) have been recorded as parasites of bats. The five genera with about 32 species of the family Polyctenidae are parasitic on bats in tropical and subtropical regions worldwide (Esberard et al. 2005; Marshall 1982). Species of the family Cimicidae (23 genera, 108 species), depending on their genus, are parasitic mainly on mammals, e.g. bats, or birds. Only one genus uses hosts in both classes of animals. The family Cimicidae occurs all over the world (Autino et al. 2009; Krinsky 2002; Weirauch and Gerry 2009). In the class Arachnida, mites and ticks of the following families also parasitise bats: Argasidae, Chirodiscidae, Ixodidae, Laelapidae, Leeuwenhoekiidae, Listrophoridae, Macronyssidae, Myobiidae, Rosensteiniidae, Spinturnicidae, Sarcoptidae, Trombiculidae and Spelaeorhynchidae.

5.2 Host–Parasite List

The list (Table 5.1) presents a comprehensive overview of the ectoparasites of the suborder Microchiroptera from South and Central America. The presented list contains the records for a total of 172 bat species of 9 families as hosts. Additionally, we added 44 host records where only genus is given. With 102 mentioned species the family Phyllostomidae represents the largest group of hosts. Followed by the families Vespertilionidae (37 species), Molossidae (15 species), Emballonuridae (7 species) and Mormoopidae (5 species). The lowest numbers of species contain the four bat families Natalidae and Noctilionidae (each 2 species) and Furipteridae and Thyropteridae (each 1 species). The different numbers of recorded hosts of the families are primarily explained by the different number of family members. The amount of associated parasite species is 273 and split up as follows: apart from 141 species of Streblidae and 46 species of Nycteribiidae (we added 102 records of both families where only genus of parasite is given), we were also able to include 8 flea species, 4 bug species (we added 2 records where only genus of parasite is given) as well as 74 species including mites and ticks (we added 32 records where only genus of parasite is given). The numerically largest proportion takes the species of the both families Streblidae and Nycteribiidae. In case of the families Streblidae and Nycteribiidae, the recorded species represent almost 61.57 %, respectively 16.08 % of the worldwide known species of these families. We included the data from the following 20 countries: Argentina, Belize, Bolivia, Brazil, Chile, Colombia, Costa Rica, Ecuador, French Guiana, Guatemala, Guyana, Honduras, Mexico, Nicaragua, Panama, Paraguay, Peru, Suriname, Uruguay and Venezuela. The data are mostly collated from primary sources to ensure that the list is as accurate as possible. In nearly all cases, we accessed works in which parasites were obtained directly from a host, e.g. parasites from roosting places were included only if parallel given in the publication. Additional information about the host distribution derived from

Table 5.1 Host parasite list

Host (occurrence)	Parasite (occurrence parasite with host)	References
Emballonuridae		
Cormura brevirostris (BO,BR, CO,CR,EC,GF,GY,NI,PA, PE,SR,VE)	*Strebla cormurae* (VE)	Sampaio et al. (2008f), Wenzel (1976)
Peropteryx kappleri (BO,BR, BZ,CO,CR,EC,GF,GT,GY, HO,MX,NI,PA,PE,SR,SV, VE)	*Strebla proxima* (VE), *Trichobius longipilis* (VE)	Dávalos et al. (2008a), Wenzel (1976)
Peropteryx macrotis (BO,BR, BZ,CO,CR,EC,GF,GT,GY, HN,MX,NI,PA,PY,PE,SR, SV,VE)	*Exastinion clovisi* (VE), *Metelasmus pseudopterus* (VE), *Ornithodoros azteci* (PA), *Strebla proxima* (VE), *Tecomatlana sandovali,* (PA), *Trichobius longipilis* (GT,VE)	Barquez et al. (2008e), Dick (2006), Wenzel (1976), Wenzel and Tipton (1966)
Peropteryx trinitatis (GF,VE)	*Strebla proxima* (VE)	Sampaio et al. (2008a), Wenzel (1976)
Saccopteryx bilineata (BO,BR, BZ,CO,CR,EC,GF,GT,GY, HN,MX,NI,PA,PE,SR,VE)	*Basilia dubia* (VE), *Beamerella acutascuta* (PA), *Euschoengastia desmodus* (PA), *Noctiliostrebla aitkeni* (VE), *Strebla alvarezi* (GT, PA), *Strebla asternalis* (VE), *Trichobius parasiticus* (VE), *Trombicula saccopteryx* (PA)	Dick (2006), Graciolli (2001), Hofstede et al. (2004), Sampaio et al. (2008f), Wenzel (1976), Wenzel and Tipton (1966)
Saccopteryx leptura (BO,BR, CO,CR,EC,GF,GT,GY,HN, MX,NI,PA,PE,SR,VE)	*Trichobius costalimai* (BR)	Dias et al. (2009)
Saccopteryx sp. (VE)	*Strebla asternalis* (VE)	Wenzel (1976)
Rhynchonycteris naso (BO,BR, BZ,CO,CR,EC,GF,GT,GY, HN,MX,NI,PA,PE,SR,VE)	*Strebla hirsutus* (VE), *Trichobius caecus* (VE), *Trichobius longipes* (VE)	Hofstede et al. (2004), Wenzel (1976)
Furipteridae		
Furipterus horrens (BR,CO, GF,GT,PE,SR,VE)	*Trichobius pallidus* (VE)	Wenzel (1976)
Molossidae		
Cynonomops (= *Molossops*) *planirostris* (PA)	*Hesperoctenes fumarius* (PA)	Wenzel and Tipton (1966)
Eumops bonariensis[a] (AR,PA)	*Hesperoctenes fumarius* (AR), *Hesperoctenes* sp. (PA)	Autino et al. (2009), Wenzel and Tipton (1966)
Eumops glaucinus (AR,BO,BR, GT,MX,PA,PE,PY,VE)	*Hesperoctenes angustatus* (PA), *Nycterophilia coxata* (VE), *Strebla christinae* (VE), *Trichobioides perspicillatus* (VE), *Trichobius jubatus* (PY)	Dick and Gettinger (2005), Wenzel and Tipton (1966), Wenzel (1976)

(continued)

Table 5.1 (continued)

Host (occurrence)	Parasite (occurrence parasite with host)	References
Eumops patagonicus (AR,BO, PY)	*Strebla diaemi* (PY), *Trichobius jubatus* (PY)	Dick and Gettinger (2005)
Eumops perotis (AR,BO,BR, CO,EC,ME,PA,PE,VE)	*Rhynchopsyllus pulex* (AR)	Autino et al. (2009), Barquez and Diaz (2008)
Molossops temminckii (AR,BO, BR,CO,EC,GF,GT,PA,PE, PY,SR,UY,VE)	*Basilia carteri* (AR,PY), *Hesperoctenes vicinus* (AR), *Ornithodoros* sp. (*hasei* complex) (AR), *Trichobius jubatus* (PY)	Autino et al. (2009), Dick and Gettinger (2005), Graciolli (2001), Nava et al. (2007)
Molossus aztecus (GT,MX,VE)	*Paradyschiria parvula* (VE), *Trichobius jubatus* (VE), *Trichobius lionycteridis* (VE)	Wenzel (1976)
Molossus coibensis (BR,PA, PE,VE)	*Chiroptonyssus venezolanus* (= *Ichoronyssus venezolanus*) (PA), *Hesperoctenes* sp. (PA), *Hormopsylla kyriophila* (PA), *Ptilopsylla dunni* (PA)	Wenzel and Tipton (1966)
Molossus currentium (= *bondae*) (AR,BR,CO,CR, EC,HN,NI,PA,PE,PY,VE)	*Trichobius dunni* (PA)	Barquez et al. (2008d), Wenzel and Tipton (1966)
Molossus molossus (AR,BO, BR,CO,CR,EC,GF,GT,GY, HN,MX,NI,PA,PE,PY,SR, UY,VE)	*Basilia carteri* (BO), *Basilia ferrisi* (VE), *Basilia speiseri* (BR), *Hesperoctenes fumarius* (AR), *Hesperoctenes* sp. (*H. fumarius*?), s. above (BR), *Neotrichobus delicates* (PE), *Trichobius jubatus* (PY)	Autino et al. (2009, 2011), Dick and Gettinger (2005), Graciolli (2001), Whitaker and Mumford (1977)
Molossus rufus (= *ater*) (AR, BO,BR,BZ,CO,CR,EC,GF, GT,GY,HN,MX,NI,PA,PY, PE,SV,UY)	*Noctiliostrebla maai* (VE), *Paradyschiria curvata* (VE), *Paradyschiria parvula* (VE), *Trichobius jubatus* (BZ,PY, VE), *Trichobius longipes* (VE)	Barquez et al. (2008d), Dick and Gettinger (2005), Hofstede et al. (2004), Wenzel (1976)
Molossus sp. (GT)	*Basilia ferrisi* (GT)	Graciolli (2001)
Molossus sp. (PA)	*Ornithodoros hasei* (PA)	Wenzel and Tipton (1966)
Nyctinomops laticaudatus (AR, BO,BR,BZ,CO,EC,GF,GT, GY,HN,MX,PA,PE,PY,SV, UR,VE)	*Paraeuctenodes longipes* (VE)	Barquez et al. (2008d), Wenzel (1976)
Promops nasutus (AR,BO,BR, EC,GY,PA,PE,SR,VE)	*Hesperoctenes fumarius* (AR)	Autino et al. (2009)
Tadarida brasiliensis (AR,BO, BR,CL,MX,PA,PE)	*Basilia carteri* (AR), *Basilia currani* (AR,BR), *Basilia forcipata* (MX), *Basilia peruvia* (PE), *Carios kelleyi* (MX), *Chiroptonyssus*	Autino et al. (2009), Graciolli (2001), Guzman-Cornejo et al. (2003), Wenzel and Tipton (1966)

(continued)

Table 5.1 (continued)

Host (occurrence)	Parasite (occurrence parasite with host)	References
	robustipes (= *Ichoronyssus robustipes*) (MX,PA), *Dentocarpus macrotrichus* (MX), *Ewingana inaequalis* (MX), *Ewingana longa* (MX), *Leptotrombidium Mexicana* (MX), *Macronyssus unidens* (MX), *Myodopsylla isidori* (AR), *Notoedres lasionycteris* (MX), *Notoedres* sp. (MX) *Nycteriglyphus bifolium* (MX), *Olabidocarpus nyctinomus* (MX), *Spinturnix* sp. (PA), *Sternopsylla distincta* (PA), *Whartonia* sp. (MX)	
Tadarida yucatanica (PA)	*Hormopsylla kyriophila* (PA), *Ptilopsylla dunni* (PA), *Rhynchopsyllus megastigmata* (PA), *Sternopsylla distincta* (PA)	Wenzel and Tipton (1966)
Mormoopidae		
Mormoops megalophylla (GT, VE)	*Nycterophilia coxata* (VE), *Nycterophilia fairchildi* (VE), *Nycterophilia mormoopsis* (GT,VE), *Trichobius caecus* (VE), *Trichobius leionotus* (GT,VE), *Trichobius parasiticus* (VE)	Dick (2006), Wenzel (1976)
Pteronotus davyi[b] (?) (BZ,CR, VE)	*Nycterophilia coxata* (VE), *Nycterophilia fairchildi* (GT[c], VE), *Nycterophilia mormoopsis* (VE), *Nycterophilia parnelli* (VE), *Trichobius caecus* (VE), *Trichobius galei* (BZ,VE), *Trichobius johnsonae* (BZ, CR,VE)	Bärtschi (2000), Dick (2006), Miller and Tschapka (2009), Wenzel (1976)
Pteronotus gymnonotus (= *suapurensis*) (BO,BR,BZ, CO,CR,EC,GT,GY,HN, MX,NI,PA,PE,VE)	*Nycterophilia coxata* (VE), *Nycterophilia fairchildi* (PA, VE), *Periglischrus elongates* (PA), *Trichobius bilobus* (VE), *Trichobius johnsonae* (PA,VE), *Trichobius yunkeri* (PA), *Trombicula tibbettsi* (PA)	Dick (2006), Wenzel (1976), Wenzel and Tipton (1966)

(continued)

Table 5.1 (continued)

Host (occurrence)	Parasite (occurrence parasite with host)	References
Pteronotus parnellii[b] (BR,BZ, CR,GT,MX,PA,VE)	*Alexfainia chilonycteris* (PA), *Aspidoptera buscki* (VE), *Nycterophilia coxata* (VE), *Nycterophilia fairchildi* (BZ, VE), *Nycterophilia natali* (VE), *Nycterophilia parnelli* (BZ,CR,GT[c],PA,VE), *Ornithodoros viguerasi* (PA), *Paradyschiria lineata* (VE), *Periglischrus elongates* (PA), *Periglischrus* sp. (PA), *Speiseria ambigua* (PA), *Strebla altmanni* (PA), *Trichobioides perspicillatus* (BZ), *Trichobius caecus* (BZ, CR,VE), *Trichobius costalimai* (PA), *Trichobius joblingi* (PA,VE), *Trichobius johnsonae* (BZ), *Trichobius parasparsus* (VE), *Trichobius sparsus* (BZ,CR,PA,VE), *Trichobius sphaeronotus* (VE), *Trichobius yunkeri* (BZ,CR,PA), *Trombicula anophthalma* (PA), *Vergrandia galei* (PA)	Bärtschi (2000), Dick (2006), Hofstede et al. (2004), Miller and Tschapka (2009), Timm et al. (1989), Timm and LaVal (1998), Wenzel (1976), Wenzel and Tipton (1966)
Pteronotus personatus[d] (PA,VE)	*Alexfainia munozi* (PA), *Amblyomma* sp. (PA), *Antricola mexicanus* (PA), *Nycterophilia fairchildi* (PA), *Perates insessus* (PA), *Trichobius johnsonae* (PA, VE), *Trombicula monops* (PA)	Wenzel (1976), Wenzel and Tipton (1966)
Pteronotus sp. (PA)	*Ornithodoros viguerasi* (PA)	Wenzel and Tipton (1966)
Natalidae		
Natalus sp. (PA)	*Trichobius galei* (PA)	Wenzel and Tipton (1966)
Natalus stramineus (BZ,GT, PA,PY)	*Nycterophilia natali* (GT[c],PA), *Mesoperiglischrus natali* (= *Periglischrus natali*) (PA), *Nycterophilia parnelli* (BZ), *Phalconomus* sp. (GT), *Speiseria ambigua* (PA), *Strebla carolliae* (PA), *Trichobius dugesioides* (PA), *Trichobius galei* (BZ,PA, PY), *Trichobius joblingi* (PA)	Bärtschi (2000), Dick (2006), Dick and Gettinger (2005), Wenzel and Tipton (1966)

(continued)

Table 5.1 (continued)

Host (occurrence)	Parasite (occurrence parasite with host)	References
Natalus tumidirostris (CO,GF, GT,GY,SR,VE)	*Nycterophilia coxata* (VE), *Nycterophilia fairchildi* (VE), *Nycterophilia natali* (GT[c], VE), *Trichobius galei* (VE), *Trichobius sparsus* (VE)	Wenzel (1976), Dick (2006)
Noctilionidae		
Noctilio albiventris (= *N. labialis*, = *Dirias labialis*) (AR,BO,BR,BZ,CR,EC, GF,GY,HN,MX,NI,PA,PE, PY,VE)	*Noctiliostrebla maai* (PA,VE), *Neotrichobus delicates* (PE), *Noctiliostrebla traubi* (VE), *Ornithodoros hasei* (BO,PA), *Paradyschiria curvata* (VE), *Paradyschiria fusca* (BO,PE), *Paradyschiria lineata* (VE), *Paradyschiria parvula* (BR, PE,PY,VE), *Paradyschiria parvuloides* (PA,VE), *Strebla curvata* (VE), *Trichobius joblingi* (PA,VE), *Trichobius johnsonae* (VE), *Trichobius parasiticus* (VE), *Xenotrichobius noctilionis* (PY,VE)	Autino et al. (2011), Dias et al. (2009), Dick and Gettinger (2005), Dick et al. (2007), Wenzel (1976), Wenzel and Tipton (1966)
Noctilio leporinus (AR,BO,BR, BZ,CO,CR,EC,GF,GT,GY, HN,MX,NI,PA,PE,PY,SR, VE)	*Latrocimex spectans* (AR), *Megistopoda aranea* (VE), *Megistopoda* sp. (VE), *Neotrichobius stenopterus* (PA), *Noctiliostrebla aitkeni* (AR,BR,PE,PY,VE), *Noctiliostrebla dubia* (AR, PY,VE), *Noctiliostrebla maai* (PY,VE), *Noctiliostrebla traubi* (BZ,PA,VE), *Ornithodoros hasei* (PA), *Paradyschiria fusca* (AR,BR, BZ,PE,PY), *Paradyschiria lineata* (PA,VE), *Paradyschiria* sp. (AR), *Periglischrus aitkeni* (PA), *Strebla guajiro* (BO), *Trichobius joblingi* (VE), *Trichobius parasiticus* (VE), *Xenotrichobius noctilionis* (AR,VE)	Autino et al. (2009, 2011), Bärtschi (2000), Dick and Gettinger (2005), Dick et al. (2007), Hofstede et al. (2004), Wenzel (1976), Wenzel and Tipton (1966), Whitaker and Mumford (1977)
Phyllostomidae		
Ametrida centurio (BR,CO,EC, GF,GY,PA,SV,VE)	*Anastrebla spurrelli* (VE)	Miller et al. (2008), Wenzel (1976)

(continued)

Table 5.1 (continued)

Host (occurrence)	Parasite (occurrence parasite with host)	References
Anoura caudifera (AR,BO,BR, CO,EC,GF,GY,PA,PE,VE)	*Anastrebla caudiferae* (AR,BO, BR,VE), *Anastrebla modestini* (BR), *Anastrebla* sp. (BR), *Aspidoptera buscki* (VE), *Exastinion clovisi* (BR, VE), *Strebla guajiro* (BO), *Trichobius* sp. (BR), *Trichobius tiptoni* (BR,BO, VE)	Autino et al. (2009), Bertola et al. (2005), Dick et al. (2007), Eriksson et al. (2011), Rui and Graciolli (2005), Wenzel (1976), Whitaker and Mumford (1977)
Anoura cultrata (BO,BR,CO, CR,EC,PA,PE,SA,VE)	*Anastrebla mattadeni* (CR, PA, VE), *Exastinion clovisi* (CR, PA), *Exastinion oculatum* (CR,VE), *Periglischrus vargasi* (PA)	Mantilla and Molinari (2008), Miller and Tschapka (2009), Timm et al. (1989), Wenzel (1976), Wenzel and Tipton (1966)
Anoura geoffroyi (BO,BR,CO, CR,EC,GF,GT,GY,HN, MX,NI,PA,PE,PY,SR,VE)	*Anastrebla caudiferae* (BR), *Anastrebla mattadeni* (CR, PA,VE), *Anastrebla modestini* (BO,BR,CR,GT, PA,PE,PY,VE), *Basilia speiseri* (BR), *Exastinion clovisi* (BR,CR,PA,PE,VE), *Exastinion deceptivum* (BO), *Periglischrus vargasi* (MX, PA), *Strebla harderi* (BR, VE), *Strebla paramirabilis* (VE), *Trichobius caecus* (VE), *Trichobius costalimai* (VE), *Trichobius joblingi* (VE), *Trichobius propinquus* (VE), *Trichobius* sp. (BR), *Trichobius tiptoni* (BR), *Trichobius uniformis* (VE)	Bertola et al. (2005), Claps et al. (2005), Dick (2006), Dick et al. (2007), Gordon and Owen (1999), Graciolli (2001), Komeno and Linhares (1999), Miller and Tschapka (2009), Sheeler-Gordon and Owen (1999), Wenzel (1976), Wenzel and Tipton (1966), Whitaker and Mumford (1977)
Anoura latidens (CO,GY,PE, VE)	*Anastrebla modestini* (VE)	Dick et al. (2007)
Anoura sp. (VE)	*Anastrebla modestini* (VE), *Exastinion clovisi* (VE), *Mastoptera* sp. (*minuta* complex) (VE), *Strebla hardeni* (VE), *Trichobius imitator* (VE), *Trichobius propinquus* (VE)	Wenzel (1976)
Artibeus (*Dermanura*[e]) *anderseni* (BO,BR,CO,EC, PE)	*Neotrichobius delicatus* (BO,PE)	Autino et al. (2011), Dick et al. (2007), Sampaio et al. (2008a)
Artibeus (*Dermanura*[e]) *aztecus* (CR,GT,HN,MX,PA,SV)	*Paratrichobius* sp. (PA), *Periglischrus iheringi* (MX)	Miller et al. (2008), Sheeler-Gordon and Owen (1999), Wenzel and Tipton (1966)

(continued)

Table 5.1 (continued)

Host (occurrence)	Parasite (occurrence parasite with host)	References
Artibeus (*Dermanura*[e]) *cinereus* (BO,BR,CO,GF, GY,PE,SR,VE)	*Aspidoptera falcata* (VE), *Neotrichobius delicatus* (BR), *Neotrichobius stenopterus* (PA), *Paratrichobius lowei* (PA,VE), *Periglischrus iheringi* (PA)	Graciolli and Aguiar (2002), Sampaio et al. (2008a), Wenzel (1976), Wenzel and Tipton (1966[d])
Artibeus fimbriatus (AR,BR, PY)	*Aspidoptera falcate* (PY), *Aspidoptera phyllostomatis* (AR,PY), *Megistopoda aranea* (AR,BR,PY), *Megistopoda proxima* (BR), *Metelasmus psuedopterus* (AR,BR,PY), *Paratrichobius longicrus* (BR), *Strebla guajiro* (PY)	Autino et al. (2009), Bertola et al. (2005), Camilotti et al. (2010), Dick and Gettinger (2005), Rui and Graciolli (2005)
Artibeus fuliginosus (VE)	*Aspidotera buscki* (VE), *Mastoptera* sp. *(minuta* complex) (VE), *Megistopoda aranea* (VE), *Megistopoda* sp. (VE), *Neotrichobius bisetosus* (VE), *Strebla guajiro* (VE), *Strebla wiedemanni* (VE), *Trichobius caecus* (VE), *Trichobius joblingi* (VE)	Wenzel (1976)
Artibeus (*Dermanura*[e]) *gnomus* (BO,BR,EC,GF,GY,PE,SR, VE)	*Neotrichobius delicatus* (PE)	Autino et al. (2011), Sampaio et al. (2008a)
Artibeus hirsutus (MX)	*Periglischrus iheringi* (MX)	Sheeler-Gordon and Owen (1999)
Artibeus intermedius: s. *A. lituratus*		
Artibeus cf. *jamaicensis* (BR)	*Aspidoptera phyllostomatis* (BR)	Graciolli and Linardi (2002)
Artibeus jamaicensis[b] (BR,BZ, CR,GT,MX,PA,PE,PY,VE)	*Aspidoptera buscki* (PA,VE), *Aspidoptera falcate* (BR, VE), *Aspidoptera phyllostomatis* (BZ,GT,PY), *Basilia rondanii* (MX), *Basilia wenzeli* (PA,PY), *Exastinion clovisi* (CR,VE), *Macronyssoides kochi* (= *Ichoronyssus kochi)* (PA), *Mastoptera* sp. (*minuta* complex) (VE), *Megistopoda aranea* (BZ,CR,PA,PE,PY, VE), *Megistopoda* sp. (*proxima* complex) (VE), *Metelasmus pseudopterus* (BZ,CR,GT,PA,PY,VE),	Autino et al. (2011), Bärtschi (2000), Dick (2006), Dick and Gettinger (2005), Graciolli (2001), Graciolli and Aguiar (2002), Hofstede et al. (2004), Miller and Tschapka (2009), Sheeler-Gordon and Owen (1999); own record, Wenzel (1976), Wenzel and Tipton (1966)

(continued)

Table 5.1 (continued)

Host (occurrence)	Parasite (occurrence parasite with host)	References
	Neotrichobius bisetosus (VE), *Neotrichobius* sp. (*delicatus* complex) (VE), *Neotrichobius stenopterus* (PA), *Nycterophilia coxata* (VE), *Paratrichobius longicrus* (BZ,PA), *Periglischrus iheringi* (MX, PA), *Spelaeorhynchus* sp. (PA), *Strebla carolliae* (PA), *Strebla guajiro* (VE), *Strebla hertigi* (PA), *Strebla mirabilis* (VE), *Strebla paramirabilis* (VE), *Strebla vespertilionis* (PA), *Strebla wiedemanni* (VE), *Trichobioides perspicillatus* (VE), *Trichobius assimilis* (VE), *Trichobius caecus* (VE), *Trichobius costalimai* (VE), *Trichobius intermedius* (GT), *Trichobius joblingi* (PA,VE), *Trichobius longipes* (PA,VE), *Trichobius parasiticus* (VE), *Trichobius uniformis* (PA)	
Artibeus lituratus (including *A. intermedius*) (BR,BZ, BO,CR,PA,PE,PY,VE)	*Anastrebla caudiferae* (BR), *Aspidoptera buscki* (PA,VE), *Aspidoptera falcate* (BR,PY), *Aspidoptera phyllostomatis* (AR,BR,PE,PY), *Mastoptera* sp. (*minuta* complex) (VE), *Megistopoda aranea* (AR, BR,BZ,PA,PE,VE), *Metelasmus pseudopterus* (PA,PE,PY,VE), *Paradyschiria longicrus* (PE, PY), *Paratrichobius longicrus* (BR,BO,CR,PA), *Paratrichobius* sp. (BR), *Periglischrus iheringi* (PA), *Periglischrus* sp. (BO), *Speiseria ambigua* (PA), *Strebla mirabilis* (VE), *Strebla wiedemanni* (VE), *Trichobius caecus* (VE), *Trichobius costalimai* (PA), *Trichobius joblingi* (BR,PA, VE), *Trichobius*	Autino et al. (2009, 2011), Bärtschi (2000), Bertola et al. (2005), Dias et al. (2009), Dick and Gettinger (2005), Graciolli and Aguiar (2002), Hofstede et al. (2004), Miller and Tschapka (2009), Sheeler-Gordon and Owen (1999); own record; Wenzel (1976), Wenzel and Tipton (1966), Whitaker and Mumford (1977)

(continued)

Table 5.1 (continued)

Host (occurrence)	Parasite (occurrence parasite with host)	References
	lonchophyllae (PA), *Trichobius parasiticus* (VE), *Trichobius* sp. (BR), *Trichobius uniformis* (VE), *Trichobius urodermae* (VE), *Trichobius vampyropis* (PA), *Trichobius yunkeri* (PA)	
Artibeus obscurus (BR,PE)	*Aspidoptera phyllostomatis* (BR, PE), *Megistopoda aranea* (BR), *Metelasmus pseudopterus* (BR), *Trichobius joblingi* (BR)	Autino et al. (2011), Bertola et al. (2005), Dias et al. (2009)
Artibeus phaeotis (*Dermanura*[e]) and/or *Artibeus* (*Dermanura*[e]) *watsoni* (according to the author a final determination is not possible) (CR)	*Neotrichobius stenopterus* (CR)	Miller and Tschapka (2009)
Artibeus (*Dermanura*[e]) *phaeotis* (BZ,MX)	*Neotrichobius stenopterus* (BZ), *Periglischrus iheringi* (BZ, MX), *Speiseria peytoni* (BZ)	Bärtschi (2000), Hofstede et al. (2004), Sheeler-Gordon and Owen (1999)
Artibeus planirostris (AR,BO, BR,PE)	*Aspidoptera phyllostomatis* (AR, BR,PE), *Megistopoda aranea* (AR,BR,BO,PE), *Metelasmus pseudopterus* (AR,BR,PE), *Paratrichobius longicrus* (BR), *Trichobius joblingi* (BR)	Autino et al. (2009, 2011), Dias et al. (2009), Eriksson et al. (2011); own record
Artibeus sp. (PA)	*Ixodes* sp. (PA)	Wenzel and Tipton (1966)
Artibeus sp. (PE)	*Metelasmus pseudopterus* (PE)	Autino et al. (2011)
Artibeus sp. (VE)	*Metelasmus pseudopterus* (VE), *Neotrichobius* sp. (*delicatus* complex) (VE)	Wenzel (1976)
Artibeus sp. (BO)	*Periglischrus iheringi* (BO)	Dick et al. (2007)
Artibeus sp. (VE)	*Strebla paramirabilis* (VE) *Trichobius assimilis* (VE)	Wenzel (1976)
Artibeus (*Dermanura*[e]) *watsoni* (BZ,CO,CR,GT,HN,MX, NI,PA)	*Neotrichobius stenopterus* (BZ); *Periglischrus iheringi* (BZ, CR)	Bärtschi (2000), Hofstede et al. (2004), Miller et al. (2008), Timm et al. (1989), Timm and LaVal (1998)
Artibeus toltecus (BZ,MX,PA)	*Leptotrombidium panamensis* (PA), *Macronyssoides kochi* (= *Ichoronyssus kochi*) (PA), *Neotrichobius* sp. (BZ), *Paratrichobius* sp. (PA), *Periglischrus iheringi* (MX)	Hofstede et al. (2004), Sheeler-Gordon and Owen (1999), Wenzel and Tipton (1966)

(continued)

Table 5.1 (continued)

Host (occurrence)	Parasite (occurrence parasite with host)	References
Carollia brevicauda[f] (BO,BR, BZ,CO,CR,GF,GY,PA,PE, SR,VE)	*Basilia speiseri* (BR), *Mastoptera minuta* (PE), *Periglischrus iheringi* (BZ), *Speiseria ambigua* (CR), *Speiseria peytoni* (BZ,VE), *Strebla alvarezi* (VE), *Strebla curvata* (PE,VE), *Strebla guajiro* (BZ,CR,PE,VE), *Strebla hertigi* (PE), *Strebla mirabilis* (BZ), *Trichobius dugesii* (VE), *Trichobius joblingi* (BR,CR,VE), *Trichobius keenani* (BZ), *Trichobius lionycteridis* (BZ), *Trichobius parasiticus* (VE), *Trichobius persimilis* (VE), *Trichobius* sp. (BZ)	Autino et al. (2011), Bärtschi (2000), Dias et al. (2009), Graciolli (2001), Hofstede et al. (2004), Miller and Tschapka (2009), Sampaio et al. (2008a), Timm and LaVal (1998), Timm et al. (1989), Wenzel (1976)
Carollia castaneae (BO,BR, CO,CR,EC,GF,GY,HN,NI, PA,SR,VE,)	*Euschoengastia desmodus* (PA), *Paraeuctenodes similis* (BO), *Radfordiella* sp.(PA), *Speiseria ambigua* (CR,PA), *Strebla carolliae* (PA), *Strebla guajiro* (CR), *Trichobius joblingi* (CR,PA, VE), *Trichobius persimilis* (VE)	Dick (2007), Miller and Tschapka (2009), Wenzel (1976), Wenzel and Tipton (1966)
Carollia perspicillata (AR,BO, BR,BZ,CO,EC,GF,GT,GY, HN,MX,NI,PA,PE,PY,SR)	*Alexfainia chilonycteris* (PA), *Anastrebla modestini* (BO, VE), *Amblyomma tapirellum* (PA), *Aspidoptera busckii* (PA), *Aspidoptera falcate* (BR,PE,VE), *Aspidoptera phyllostomatis* (AR,BR), *Beamerella acutascuta* (PA), *Euschoengastia megastyrax* (PA), *Mastoptera guimaraesi* (PA), *Mastoptera minuta* (BR), *Mastoptera* sp. (*minuta* complex) (VE), *Megistopoda aranea* (BR,PA,VE), *Megistopoda proxima* (BR), *Megistopoda* sp. (*proxima* complex) (VE), *Metelasmus pseudopterus* (PA), *Nycterophilia parnelli* (PA), *Ornithodoros brodyi* (PA), *Paraeuctenodes similis* (BR, VE), *Paratrichobius* (?)	Autino et al. (2009), Bärtschi (2000), Bertola et al. (2005), Claps et al. (2005), Dias et al. (2009), Dick (2006), Dick and Gettinger (2005), Dick et al. (2007), Eriksson et al. (2011), Graciolli and Aguiar (2002), Hofstede et al. (2004), Lourenço and Esbérard (2011); own record, Timm and LaVal (1998), Timm et al. (1989), Wenzel (1976), Wenzel and Tipton (1966)

(continued)

Table 5.1 (continued)

Host (occurrence)	Parasite (occurrence parasite with host)	References
	longicrus (PE,VE), *Paratrichobius dunni* (PA), *Periglischrus* sp. (PA), *Radfordiella* sp. (PA), *Radfordiella* sp. (PA), *Speiseria ambigua* (BR,BZ, BO,CR,PA,PE,VE), *Spelaeorhynchus* sp. (PA), *Strebla altmani* (PA), *Strebla carolliae* (PA), *Strebla christinae* (VE), *Strebla curvata* (VE), *Strebla guajiro* (BO,BR,BZ,CR,GT,PE,PY, VE), *Strebla hertigi* (VE), *Strebla matsoni* (VE), *Strebla mirabilis* (PA,VE), *Strebla vespertilionis* (PA), *Strebla wiedemanni* (VE), *Tecomatlana vesperugini* (= *Hooperella vesperuginis*,= *Trombicula vesperuginis*) (PA), *Trichobioides perspicillatus* (VE), *Trichobius caecus* (VE), *Trichobius costalimai* (BR, PA,VE), *Trichobius dugesii* (PA), *Trichobius dugesioides* (PA), *Trichobius joblingi* (BO,BR,CR,PA,PY,VE), *Trichobius johnsonae* (PA), *Trichobius keenani* (VE), *Trichobius lionycteridis* (VE), *Trichobius longipes* (PA), *Trichobius macrophylli* (PA), *Trichobius parasiticus* (VE), *Trichobius parasparsus* (VE), *Trichobius* sp. (BR), *Trichobius* sp. (BZ), *Trichobius sparsus* (PA), *Trichobius tiptoni* (BR, VE), *Trichobius uniformis* (BR,VE), *Trichobius urodermae* (PA), *Trichobius yunkeri* (PA)	
Carollia sp. (VE)	*Speiseria ambigua* (VE), *Strebla guajiro* (VE), *Trichobioides perspicillatus* (VE), *Trichobius costalimai* (VE), *Trichobius dugesioides* (VE), *Trichobius joblingi* (VE)	Wenzel (1976)

(continued)

Table 5.1 (continued)

Host (occurrence)	Parasite (occurrence parasite with host)	References
Carollia sp. (*castanea* complex) (PE)	*Strebla guajiro* (PE)	Autino et al. (2011)
Carollia subrufa (PA)	*Euschoengastia colombiae* (PA), *Euschoengastia desmodus* (PA), *Speiseria ambigua* (PA), *Spelaeorhynchus* sp. (PA), *Strebla carolliae* (PA), *Trichobius dugesioides* (PA), *Trichobius joblingi* (PA)	Wenzel and Tipton (1966)
Centurio senex (BO,BR,BZ, CR,GT,HN,MX,NI,PA,PE, SV,VE)	*Basilia* sp. (CR)	Graciolli (2001), Miller et al. (2008)
Chiroderma salvini (BO,BR, BZ,CR,EC,GT,HN,MX,NI, PA,PE,VE)	*Amblyomma* sp. (PA), *Paratrichobius salvini* (PA, VE), *Periglischrus iheringi* (MX,PA), *Trichobius persimilis* (VE)	Aguirre et al. (2008), Sheeler-Gordon and Owen (1999), Wenzel (1976), Wenzel and Tipton (1966)
Chiroderma trinitatum (BO, BR,CO,EC,GF,GY,PA,PE, SR,VE)	*Paratrichobius* sp. (*salvini* complex) (VE)	Miller et al. (2008), Wenzel (1976)
Chiroderma villosum (BO,BR, BZ,CO,CR,EC,GF,GT,GY, HN,MX,NI,PA,PE,SR,VE)	*Aspidoptera busckii* (PA), *Metelasmus pseudopterus* (VE), *Nycterophilia coxata* (VE), *Paratrichobius* sp. (PA), *Paratrichobius* sp. (*salvini* complex) (VE), *Strebla chrotopteri* (VE), *Strebla wiedemanni* (VE), *Trichobius dugesioides* (VE), *Trichobius joblingi* (PA), *Trichobius parasiticus* (VE)	Sampaio et al. (2008f), Wenzel and Tipton (1966), Wenzel (1976)
Choeroniscus sp. (BO)	*Anastrebla caudiferae* (BO)	Dick et al. (2007)
Choeroniscus sp. (BO)	*Strebla curvata* (BO)	Dick et al. (2007)
Chrotopterus auritus (AR,BO, BR,BZ,CO,CR,EC,GF,GT, GY,HN,MX,NI,PA,PE,SR, VE)	*Basilia hughscotti* (BR), *Basilia ortizi* (BR), *Craneopsylla minerva* (AR), *Strebla chrotopteri* (AR,BR,PY,VE), *Strebla mirabilis* (VE), *Trichobius dugesioides* (BR, PA,VE)	Autino et al. (2009), Dick and Gettinger (2005), Eriksson et al. (2011), Graciolli (2004), Graciolli et al. (2006), Wenzel (1976), Wenzel and Tipton (1966)
Desmodus rotundus (AR,BO, BR,BZ,CL,CO,CR,EC,GT, HN,MX,NI,PA,PE,PY,UY, VE)	*Aspidoptera falcata* (BR,PE), *Aspidoptera phyllostomatis* (AR,BR), *Basilia ferrisi* (VE), *Mastoptera* sp. (*minuta* complex) (VE), *Megistopoda aranea* (AR,PA,PE,VE), *Ornithodoros azteci* (PA),	Aguiar and Antonini (2011), Autino et al. (2009), Barquez et al. (2008c), Bertola et al. (2005), Claps et al. (2005), Dick and Gettinger (2005), Dick (2006),

(continued)

Table 5.1 (continued)

Host (occurrence)	Parasite (occurrence parasite with host)	References
	Ornithodoros brodyi (PA), *Paradyschiria curvata* (VE), *Paratrichobius dunni* (VE), *Paratrichobius longicrus* (BR), *Paratrichobius* sp. (*longicrus* complex) (VE), *Periglischrus desmodi* (PA), *Periglischrus herrerai* (MX), *Periglischrus iheringi* (MX, PA), *Radfordiella* sp. (PA), *Speiseria ambigua* (PA), *Strebla consocius* (VE), *Strebla diphyllae* (GT), *Strebla guajiro* (VE), *Strebla hertigi* (PA), *Strebla mirabilis* (BR), *Strebla vespertilionis* (PA), *Strebla wiedemanni* (BO,BR,BZ,CR,PE,PY,VE), *Trichobioides perspicillatus* (PA), *Trichobius caecus* (VE), *Trichobius costalimai* (PA,VE), *Trichobius dugesioides* (BR,PA,VE), *Trichobius furmani* (BR), *Trichobius joblingi* (BR,PA, VE), *Trichobius longipes* (PY,VE), *Trichobius parasiticus* (AR,BR,BZ,CR, PA,PY), *Trichobius tiptoni* (VE), *Trichobius uniformis* (PA)	Eriksson et al. (2011), Graciolli et al. (2006), Graciolli (2001), Graciolli and Linardi (2002), Hofstede et al. (2004); own record, Sheeler-Gordon and Owen (1999), Timm and LaVal (1998), Timm et al. (1989), Wenzel (1976), Wenzel and Tipton (1966)
Diaemus (*Desmodus*) *youngii* (AR,BO,BR,CO,CR,EC, GF,GT,GY,MX,NI,PA,PE, PY,SR,VE)	*Radfordiella oudemans* (BO), *Strebla diaemi* (BR,PA,PY, VE), *Trichobius diaemi* (BR, VE), *Trichobius parasiticus* (PA,VE)	Barquez et al. (2008c), Dias et al. (2009), Dick and Gettinger (2005), Dick et al. (2007), Wenzel (1976), Wenzel and Tipton (1966)
Diphylla ecaudata (BO,BR,BZ, CO,CR,EC,GT,HN,MX,NI, PA,PE,VE)	*Strebla diphyllae* (GT), *Strebla mirabilis* (BR), *Strebla wiedemanni* (BR), *Trichobius diphyllae* (VE), *Trichobius furmani* (BR), *Trichobius parasiticus* (BR), *Trichobius uniformis* (BR)	Aguiar and Antonini (2011), Dick (2006), Sampaio et al. (2008a), Wenzel (1976)
Enchisthenes (= *Artibeus*) *hartii* (BO,BR,CO,CR,EC, GT,HN,MX,PA,PE,VE)	*Eudusbabekia paralepidoseta* (MX), *Paratrichobius sanchezi* (PA,VE), *Periglischrus iheringi* (PA), *Basilia ortizi* (VE)	Graciolli (2001), Morales-Malacara et al. (2011), Wenzel (1976), Wenzel and Tipton (1966)

(continued)

Table 5.1 (continued)

Host (occurrence)	Parasite (occurrence parasite with host)	References
Glossophaga commissarisi (BR,BZ,CO,CR,EC,GT, GY,HN,MX,NI,PA,PE)	*Anastrebla nycteridis* (CR), *Eldunnia breviceps* (CR), *Paratrichobius* sp. (*longicrus*?) (CR), *Speiseria ambigua* (CR), *Strebla curvata* (PE), *Strebla guajiro* (CR), *Trichobius joblingi* (CR)	Autino et al. (2011), Miller and Tschapka (2009)
Glossophaga leachii (GT,MX)	*Periglischrus caligus* (MX), *Strebla guajiro* (GT)	Dick (2006), Sheeler-Gordon and Owen (1999)
Glossophaga longirostris (BR, CO,GY,VE)	*Megistopoda aranea* (VE), *Nycterophilia coxata* (VE), *Paraeuctenodes longipes* (VE), *Strebla curvata* (VE), *Strebla wiedemanni* (VE), *Trichobioides perspicillatus* (VE), *Trichobius dugesii* (VE), *Trichobius parasiticus* (VE), *Trichobius sphaeronotus* (VE), *Trichobius uniformis* (VE)	Wenzel (1976)
Glossophaga morenoi (MX)	*Periglischrus caligus* (MX)	Sheeler-Gordon and Owen (1999)
Glossophaga soricina (AR,BO, BR,BZ,CO,CR,GF,GY,HN, MX,NI,PA,PE,PY,SR,VE)	*Anastrebla caudiferae* (BR), *Aspidoptera buscki* (BR,VE), *Aspidoptera falcate* (BR), *Eldunnia breviceps* (PA), *Euschoengastia desmodus* (PA), *Mastoptera minuta* (BR), *Megistopoda aranea* (BR), *Megistopoda proxima* (BR), *Paraeuctenodes longipes* (BR,VE), *Paratrichobius* sp. (*longicrus* complex) (VE), *Periglischrus caligus* (MX,PA), *Speiseria ambigua* (BR,PA,PY), *Spelaeorhynchus* sp. (PA), *Strebla alvarezi* (VE), *Strebla carolliae* (PA), *Strebla curvata* (BR,BZ,PY,VE), *Strebla guajiro* (VE,PE), *Strebla kohlsi* (PE), *Strebla wiedemanni* (VE), *Tecomatlana vesperuginis* (= *Hooperella vesperuginis*,= *Trombicula vesperuginis*) (PA), *Trichobius dugesii* (BR, BZ,PA,PY,VE), *Trichobius*	Autino et al. (2009, 2011), Bertola et al. (2005), Dias et al. (2009), Dick and Gettinger (2005), Eriksson et al. (2011), Sheeler-Gordon and Owen (1999), Graciolli and Aguiar (2002), Hofstede et al. (2004), Graciolli and Rui (2001), Sheeler-Gordon and Owen (1999); own record, Wenzel (1976), Wenzel and Tipton (1966)

(continued)

Table 5.1 (continued)

Host (occurrence)	Parasite (occurrence parasite with host)	References
	joblingi (BO,BR,PA), *Trichobius parasiticus* (VE), *Trichobius propinquus* (VE), *Trichobius* sp. (BR), *Trichobius uniformis* (BR, BZ,PA,PY,VE)	
Glossophaga sp. (BO)	*Periglischrus caligus* (BO)	Dick et al. (2007)
Glyphonycteris (= *Micronycteris*) *daviesi* (BO, BR,CO,CR,EC,GF,GU,HN, PA,PE,SR,VE)	*Trichobius yunkeri* (CR)	Ochoa and Velazco (2008), Timm et al. (1989)
Glyphonycteris (= *Micronycteris*) *sylvestris* (BO,BR,CO,CR,GF,GY, HN,MX,NI,PA,PE,SR,VE)	*Strebla alvarezi* (CR,PA)	Miller and Tschapka (2009), Wenzel and Tipton (1966)
Hylonycteris underwoodi (BZ, CR,GT,HN,MX,NI,PA)	*Basilia rondanii* (MX), *Strebla* sp. (CR)	Graciolli (2001), Miller and Tschapka (2009)
Lampronycteris (= *Micronycteris*) *brachyotis* (BR,BZ,CO,CR,GF,GT, GY,MX,NI,PA,PE,SR,VE)	*Trichobius joblingi* (GT), *Trichobius lionycteridis* (BZ), *Trichobius tuttlei* (VE)	Bärtschi (2000), Dick (2006), Wenzel (1976)
Leptonycteris curasoae (CO, MX,VE)	*Megistopoda* sp. (*proxima* complex) (VE), *Nycterophilia coxata* (VE), *Nycterophilia fairchildi* (VE), *Periglischrus paracaligus* (MX), *Trichobius caecus* (VE), *Trichobius galei* (VE), *Trichobius parasiticus* (VE), *Trichobius sphaeronotus* (VE)	Sheeler-Gordon and Owen (1999), Wenzel (1976)
Leptonycteris nivalis (MX,GT)	*Basilia antrozi* (MX), *Eudusbabekia provirilia* (MX), *Periglischrus paracaligus* (MX)	Graciolli (2001), Morales-Malacara et al. (2002), Sheeler-Gordon and Owen (1999)
Lichonycteris obscura (BO,BR, BZ,CO,CR,EC,GF,GT,GY, HN,MX,NI,PA,PE,SR,VE)	*Anastrebla modestini* (CR), *Strebla* sp. (CR)	Miller and Tschapka (2009)
Lionycteris spurrelli (BO,BR, CO,EC,GF,GY,PA,PE,SR, VE)	*Anastrebla spurrelli* (VE), *Trichobius lionycteridis* (PA, VE)	Wenzel (1976), Wenzel and Tipton (1966)
Lonchophylla dekeyseri (BR, CO,CR,EC,NI,PA,PE,VE)	*Trichobius lonchophyllae* (BR)	Dávalos et al. (2008a), Graciolli and Aguiar (2002)
Lonchophylla robusta (CO,CR, EC,NI,PA,PE,VE)	*Anastrebla modestini* (CR), *Anastrebla nycteridis* (CR, PA,VE), *Anatrichobius*	Dávalos et al. (2008b), Miller and Tschapka (2009), Timm et al. (1989),

(continued)

Table 5.1 (continued)

Host (occurrence)	Parasite (occurrence parasite with host)	References
	scorzai (VE), *Eldunnia breviceps* (CR,PA), *Periglischrus* sp. (PA), *Phalconomus puliciformis* (VE), *Speiseria ambigua* (CR, PA), *Strebla carolliae* (PA), *Trichobius joblingi* (PA), *Trichobius johnsonae* (PA), *Trichobius lonchophyllae* (CR,PA,VE)	Wenzel (1976), Wenzel and Tipton (1966)
Lonchophylla thomasi (VE)	*Strebla alvarezi* (VE)	Wenzel (1976)
Lonchorhina aurita (BO,BR, BZ,CO,CR,EC,GF,GT,GY, HN,MX,NI,PA,PE,SR,SV, VE)	*Basilia wenzeli* (VE), *Megistopoda aranea* (VE), *Megistopoda* sp. (VE), *Nycterophilia parnelli* (VE), *Ornithodoros azteci* (PA), *Speiseria ambigua* (CR,PA, VE), *Strebla altmani* (BR,PA, VE), *Strebla carolliae* (PA), *Trichobius dugesioides* (PA), *Trichobius flagellates* (BR, CR,VE), *Trichobius joblingi* (PA), *Trichobius macrophylli* (PA), *Trichobius parasiticus* (VE), *Trichobius yunkeri* (PA)	Graciolli (2001), Graciolli and Linardi (2002), Sampaio et al. (2008f), Timm and LaVal (1998), Timm et al. (1989), Wenzel and Tipton (1966), Wenzel (1976)
Lonchorhina orinocensis (CO, VE)	*Nycterophilia parnelli* (VE), *Strebla altmani* (VE), *Trichobius ethophallus* (VE), *Trichobius flagellatus* (VE), *Trichobius parasiticus* (VE)	Ochoa and Molinari (2008), Wenzel (1976)
Lonchorhina sp. (PY)	*Basilia tiptoni* (PY)	Graciolli (2001)
Lophostoma brasiliense (= *Tonatia brasiliense*,= *Tonatia minuta*) (BO,BR, BZ,CO,CR,EC,GF,GT,GY, HN,MX,NI,PA,PE,PY,SR, VE)	*Mastoptera minuta* (BO,BR,PA, PY), *Mastoptera* sp. (*minuta* complex) (CR), *Pseudostrebla greenwelli* (PA,VE), *Strebla consocia* (PE), *Strebla hoogstraali* (CR,PA), *Strebla tonatiae* (BR,VE), *Trichobius affinis* (VE), *Trichobius mendezi* (PA), *Trichobius silvicolae* (CR)	Autino et al. (2011), Dias et al. (2009), Dick and Gettinger (2005), Miller and Tschapka (2009); own record, Sampaio et al. (2008f), Timm and LaVal (1998), Timm et al. (1989), Wenzel (1976), Wenzel and Tipton (1966)
Lophostoma carrikeri (BO,BR, CO,GF,GY,PE,SR,VE)	*Mastoptera minuta* (BR), *Mastoptera sp.* (*minuta* complex) (VE), *Pseudostrebla sparsisetis* (VE), *Stizostrebla longirostris* (BR,VE), *Strebla tonatiae* (VE), *Trichobius* sp. (VE)	Dias et al. (2009), Wenzel (1976)

(continued)

Table 5.1 (continued)

Host (occurrence)	Parasite (occurrence parasite with host)	References
Lophostoma (= *Tonatia*) *evotis* (BZ,GT,MX,HN)	*Mastoptera minuta* (GT), *Paradyschiria* sp. (BZ)	Dick (2006), Hofstede et al. (2004)
Lophostoma silvicolum (= *Tonatia silvicola*) (BO,BR, CO,CR,EC,GF,GY,HN,NI, PA,PE,PY,SV,SY,VE)	*Basilia constricta* (EC), *Mastoptera minuta* (BR,CR, PA,PE,VE), *Ornithodoros hasei* (PA), *Pseudostrebla ribeiroi* (PA,PE,VE), *Strebla consocia* (PE), *Strebla guajiro* (VE), *Strebla kohlsi* (PA,PE,VE), *Strebla mirabilis* (CR), *Strebla paramirabilis* (PE), *Trichobius dugesioides* (VE), *Trichobius dybasi* (PA), *Trichobius silvicolae* (CR, VE)	Autino et al. (2011), Barquez et al. (2008a), Dias et al. (2009), Graciolli (2001), Miller and Tschapka (2009), Wenzel (1976), Wenzel and Tipton (1966)
Macrophyllum macrophyllum (AR,BO,BR,BZ,CO,CR, EC,GF,GT,GY,HN,MX,NI, PA,PE,PY,SR,VE)	*Basilia constricta* (VE), *Periglischrus* sp. (PA), *Strebla altmani* (PA,VE), *Strebla carolliae* (PA), *Strebla matsoni* (VE), *Trichobius caecus* (VE), *Trichobius dugesioides* (VE), *Trichobius joblingi* (PA,VE), *Trichobius macrophylli* (CR, PA)	Graciolli (2001), Rodriguez and Pineda (2008), Timm et al. (1989), Timm and LaVal (1998), Wenzel (1976), Wenzel and Tipton (1966)
Macrotus waterhousii (BZ,GT, MX)	*Periglischrus delfinadoae* (MX)	Sheeler-Gordon and Owen (1999)
Mesophylla ectophylla (PE)	*Neotrichobius ectophyllae* (PE)	Autino et al. (2011)
Mesophylla[c] *macconnelli* (BO, BR,CO,CR,EC,GF,GY,NI, PA,PE,SR,VE)	*Neotrichobius ectophyllae* (VE), *Neotrichobius* sp. (CR)	Miller and Tschapka (2009), Sampaio et al. (2008c), Wenzel (1976)
Micronycteris hirsuta (BR,CO, CR,EC,GF,GY,HN,NI,PA, PE,SR,VE)	*Strebla alvarezi* (CR)	Miller and Tschapka (2009)
Micronycteris megalotis (BR, CO,PE,VE)	*Basilia bequaerti* (CO), *Speiseria ambigua* (VE), *Strebla alvarezi* (VE), *Strebla tonatiae* (PE), *Trichobius joblingi* (BR), *Trichobius keenani* (VE), *Trichobius* sp. (VE)	Autino et al. (2011), Bertola et al. (2005), Graciolli (2001), Wenzel (1976)
Micronycteris microtis (BR, BZ,PA,VE)	*Beamerella acutascuta* (PA), *Euschoengastia desmodus* (PA), *Periglischrus micronycteridis* (PA), *Perissopalla precaria* (PA), *Strebla alvarezi* (PA,VE), *Trichobius joblingi* (PA),	Graciolli et al. (2006), Hofstede et al. (2004), Wenzel (1976), Wenzel and Tipton (1966)

(continued)

Table 5.1 (continued)

Host (occurrence)	Parasite (occurrence parasite with host)	References
	Trichobius keenani (PA,VE), *Trichobius* sp. (*dugesii* complex) (BR)	
Micronycteris minuta (BO,BR, CO,CR,EC,GF,GY,HN,NI, PA,PE,SR,VE)	*Mastoptera minuta* (BR), *Periglischrus micronycteridis* (PA), *Stizostrebla longirostris* (BR), *Strebla machadoi* (PA,VE), *Strebla* sp. (BR), *Trichobius dugesioides* (PA), *Trichobius handleyi* (CR), *Trichobius joblingi* (VE)	Autino et al. (2011), Dias et al. (2009), Miller and Tschapka (2009), Wenzel (1976), Wenzel and Tipton (1966)
Micronycteris schmidtorum (BO,BR,BZ,CO,CR,GF, GT,HN,MX,NI,PA,PE,VE)	*Strebla machadoi* (VE), *Trichobius handleyi* CR), *Trichobius* sp. (CR)	Miller and Tschapka (2009), Sampaio et al. (2008a), Timm and LaVal (1998), Timm et al. (1989), Wenzel (1976)
Mimon bennettii (BR,CO,GF, GY,SR,VE)	*Basilia lindolphoi* (BR)	Graciolli (2004)
Mimon cozumelae (BZ,CO,CR, GT,HN,MX,NI,PA)	*Mastoptera minuta* (BZ)	Hofstede et al. (2004)
Mimon crenulatum (BO,BR, BZ,CO,CR,EC,GF,GT,GY, HN,MX,NI,PA,PE,SR,VE)	*Basilia mimoni* (BR,PE), *Basilia tiptoni* (BR,CR,PA,PY,VE), *Strebla machadoi* (PE), *Trichobius parasiticus* (VE)	Autino et al. (2011), Dias et al. (2009), Graciolli (2001), Hofstede et al. (2004), Komeno and Linhares (1999), Sampaio et al. (2008f), Timm et al. (1989), Wenzel (1976), Wenzel and Tipton (1966)
Phylloderma stenops (BO,BR, BZ,CO,CR,EC,GF,GT,GY, HN,MX,PA,PE,SR,VE)	*Megistopoda aranea* (VE), *Megistopoda* sp. (*proxima* complex) (VE), *Speiseria ambigua* (VE), *Strebla christinae* (CR,PA), *Strebla kohlsi* (PE)	Autino et al. (2011), Sampaio et al. (2008b), Timm and LaVal (1998), Timm et al. (1989), Wenzel (1976)
Phyllostomus discolor (BR,BZ, CO,CR,GF,GT,GY,HN, MX,NI,PA,PE,VE)	*Aspidoptera busckii* (PA), *Mastoptera guimaraesi* (VE), *Megistopoda aranea* (PA), *Metelasmus pseudopterus* (VE), *Strebla chrotopteri* (VE), *Strebla consocius* (VE), *Strebla hertigi* (BR,CR,PA, PE,VE), *Strebla mirabilis* (PA), *Strebla wiedemanni* (VE), *Trichobioides perspicillatus* (BR,CR,PA, VE), *Trichobius costalimai* (BR,BZ,CR,PA,VE), *Trichobius dugesioides*	Autino et al. (2011), Dias et al. (2009), Hofstede et al. (2004), Miller and Tschapka (2009), Wenzel (1976), Wenzel and Tipton (1966)

(continued)

Table 5.1 (continued)

Host (occurrence)	Parasite (occurrence parasite with host)	References
	(BR,VE), *Trichobius longipes* (PA), *Trichobius persimilis* (VE)	
Phyllostomus elongates (BR, BO,CO,EC,GF,GY,PE,SR, VE)	*Mastoptera minuta* (PE), *Mastoptera* sp. (*minuta* complex) (VE), *Nycterophilia coxata* (VE), *Strebla consocia* (BR,VE), *Strebla guajiro* (BO,VE), *Strebla hertigi* (PE), *Strebla mirabilis* (VE), *Strebla obtuse* (VE), *Trichobius dugesioides* (BR, VE), *Trichobius handleyi* (VE), *Trichobius joblingi* (BR,BO,VE), *Trichobius longipes* (VE), *Trichobius persimilis* (VE)	Autino et al. (2011), Graciolli and Aguiar (2002), Graciolli and Linardi (2002); own record, Wenzel (1976)
Phyllostomus hastatus (BO,BR, CO,CR,GF,GT,GY,HN,NI, PA,PE,PY,SR,VE)	*Aspidoptera buscki* (VE), *Aspidoptera falcate* (VE), *Blankaartia sinnamaryi* (PA), *Mastoptera guimaraesi* (PA, PY,VE), *Mastoptera minuta* (BR,PE), *Mastoptera* sp. (*minuta* complex) (VE), *Megistopoda aranea* (VE), *Megistopoda* sp. (*proxima* complex) (VE), *Metelasmus pseudopterus* (VE), *Microtrombicula carmenae* (= *Trombicula carmenae*) (PA), *Paratrichobius* sp. (*salvini* complex) (VE), *Periglischrus inflatiseta* (PA), *Periglischrus tiptoni* (PA), *Speiseria ambigua* (PA), *Strebla carolliae* (PA), *Strebla consocius* (BR,PE, VE), *Strebla guajiro* (VE), *Strebla hertigi* (CR,PA), *Strebla kohlsi* (*mirabilis* complex) (PE), *Strebla mirabilis* (PA,VE), *Trichobius costalimai* (CR), *Trichobius dugesioides* (VE), *Trichobius joblingi* (PA,VE), *Trichobius longipes* (BR,PA, PY,VE), *Trichobius parasiticus* (PA), *Trichobius silvicolae* (VE), *Trichobius yunkeri* (PA)	Autino et al. (2011), Dias et al. (2009), Dick and Gettinger (2005), Miller and Tschapka (2009), Wenzel (1976), Wenzel and Tipton (1966), Whitaker and Mumford (1977)

(continued)

Table 5.1 (continued)

Host (occurrence)	Parasite (occurrence parasite with host)	References
Phyllostomus sp. (AR)	*Basilia speiseri* (AR,BR)	Graciolli (2001)
Phyllostomus sp. (PA)	*Mastoptera guimaraesi* (PA), *Mastoptera minuta* (PA), *Trichobius longipes* (PA)	Wenzel and Tipton (1966)
Phyllostomus sp. (GT)	*Strebla mirabilis* (GT)	Dick (2006)
Platyrrhinus aurarius (GY,SR, VE)	*Paratrichobius* sp. (*longicrus* complex) (VE), *Strebla paramirabilis* (VE), *Trichobius angulatus* (VE), *Trichobius assimilis* (VE)	Sampaio et al. (2008e), Wenzel (1976)
Platyrrhinus brachycephalus (BO,BR,CO,EC,GF,GY, PE,SR,VE)	*Megistopoda aranea* (PE), *Paratrichobius dunni* (PE), *Paratrichobius* sp. (*salvini* complex) (VE), *Strebla consocia* (PE)	Autino et al. (2011), Sampaio et al. (2008a), Wenzel (1976)
Platyrrhinus (= *Vampyrops*) *helleri* (BO,BR,BZ,CO,CR, EC,GF,GT,GY,HN,MX,NI, PA,PE,SR,VE)	*Amblyomma cajennense* (PA), *Amblyomma* sp. (PA), *Mastoptera* sp. (*minuta* complex) (VE), *Megistopoda* sp. (VE), *Neotrichobius* sp. (*delicatus* complex) (VE), *Nycterophilia fairchildi* (VE), *Ornithodoros hasei* (PA), *Paratrichobius* sp. (PA), *Paratrichobius* sp. (BZ), *Paratrichobius* sp. (*salvini*?) (VE), *Periglischrus iheringi* (PA), *Strebla consocius* (VE), *Trichobius dugesii* (VE), *Trichobius lionycteridis* (VE), *Trichobius tiptoni* (VE), *Trichobius uniformis* (VE)	Hofstede et al. (2004), Miller et al. (2008), Wenzel (1976), Wenzel and Tipton (1966)
Platyrrhinus (= Vampyrops) lineatus (BO,BR,CO,EC, PA,PY,SR,UY,VE)	*Aspidoptera falcate* (PY), *Paradyschiria salvini* (PY), *Paratrichobius* aff. *Longicrus* (BR), *Paratrichobius longicrus* (BR), *Paratrichobius* sp. (BR), *Trichobius angulatus* (BR, PY), *Trichobius* sp. (BR)	Dick and Gettinger (2005), Eriksson et al. (2011), Graciolli and Aguiar (2002), Whitaker and Mumford (1977)
Platyrrhinus umbratus (CO, VE)	*Aspidoptera falcate* (VE), *Megistopoda* sp. (VE), *Paratrichobius* sp. (*longicrus* complex) (VE), *Strebla wiedemanni* (VE), *Trichobius*	Sampaio et al. (2008a), Wenzel (1976)

(continued)

Table 5.1 (continued)

Host (occurrence)	Parasite (occurrence parasite with host)	References
	parasiticus (VE), *Trichobius persimilis* (VE), *Trichobius vampyropis* (VE)	
Platyrrhinus (= *Vampyrops*) *vittatus*[a] (CO,CR,PA,VE)	*Ichoronyssus* sp. (PA), *Paratrichobius* sp. (PA), *Paratrichobius* sp. (*longicrus?*) (VE), *Periglischrus iheringi* (PA), *Speiseria ambigua* (PA), *Trichobius vampyropis* (PA, VE)	Wenzel (1976), Wenzel and Tipton (1966), Velazco et al. (2008), (Velazco 2005[a])
Rhinophylla fischerae (BO,BR, CO,EC,PE,VE)	*Neotrichobius delicatus* (PE)	Autino et al. (2011), Sampaio et al. (2008a)
Rhinophylla pumilio (BO,BR, CO,EC,GF,GU,PE,SR,VE, PE)	*Aspidoptera falcata* (PE), *Neotrichobius* sp. (*delicatus* complex) (VE), *Trichobius joblingi* (VE)	Autino et al. (2011), Sampaio et al. (2008a), Wenzel (1976)
Sphaeronycteris toxophyllum (VE)	*Aspidoptera falcata* (VE), *Trichobiu costalimai* (VE), *Trichobius dugesioides* (VE)	Wenzel (1976)
Sturnira bidens (CO,EC,PE, VE)	*Trichobius hispidus* (VE)	Pacheco et al. (2008a), Wenzel (1976)
Sturnira bogotensis (VE)	*Trichobius petersoni* (VE)	Wenzel (1976)
Sturnira erythromos (AR,BO, BR,CO,EC,PE,VE)	*Aspidoptera phyllostomatis* (AR), *Megistopoda proxima* (AR), *Megistopoda* sp. (*proxima* complex) (VE), *Periglischrus ojasti* (BO), *Trichobius joblingi* (VE), *Trichobius petersoni* (BO, VE)	Autino et al. (2009), Barquez and Diaz (2008), Dick et al. (2007), Wenzel (1976)
Sturnira lilium (AR,BO,BR, BZ,CO,CR,GT,HN,MX, PA,PE,PY,UY,VE)	*Aspidoptera delatorrei* (BO,BZ, CR,GT,PA), *Aspidoptera falcate* (BR,BZ,PE,PY,VE), *Aspidoptera phyllostomatis* (AR), *Mastoptera* sp. (*minuta* complex) (VE), *Megistopoda aranea* (VE), *Megistopoda proxima* (AR,BR,BO,BZ,CR, PA,PY,PE), *Megistopoda* sp. (*proxima* complex) (VE), *Megistopoda theodori* (BO), *Metelasmus pseudopterus* (GT), *Metelasmus wenzeli* (BR), *Paradyschiria parvuloides* (PA), *Paraeuctenodes similis* (BR), *Paratrichobius longicrus*	Autino et al. (2009), Autino et al. (2011), Bärtschi (2000), Barquez et al. (2008c), Bertola et al. (2005), Dick (2006), Dick and Gettinger (2005), Dick et al. (2007), Eriksson et al. (2011), Graciolli and Aguiar (2002), Graciolli and Bianconi (2007), Graciolli and Rui (2001), Miller and Tschapka (2009); own record, Sheeler-Gordon and Owen (1999), Wenzel (1976), Wenzel and Tipton (1966)

(continued)

Table 5.1 (continued)

Host (occurrence)	Parasite (occurrence parasite with host)	References
	(BR), *Periglischrus aitkeni* (BO,PA), *Periglischrus ojastii* (MX), *Speiseria ambigua* (VE), *Strebla galindoi* (VE), *Strebla guajiro* (VE), *Strebla hertigi* (VE), *Strebla tonatiae* (VE), *Strebla wiedemanni* (VE), *Trichobioides perspicillatus* (PA,VE), *Trichobius costalimai* (VE), *Trichobius dugesioides* (VE), *Trichobius furmani* (BR), *Trichobius joblingi* (BR,VE), *Trichobius lionycteridis* (VE), *Trichobius lonchophyllae* (VE), *Trichobius parasiticus* (VE), *Trichobius parasparsus* (VE), *Trichobius phyllostomae* (BR), *Trichobius* sp. (BR), *Trichobius uniformis* (BR)	
Sturnira ludovici (CO,CR,EC, GT,GY,HN,MX,NI,PA,PY, SV,VE)	*Aspidoptera falcate* (VE), *Megistopoda aranea* (VE), *Megistopoda* sp. (*proxima* complex)(VE), *Megistopoda theodori* (PA,PY,VE), *Metelasmus* sp. (VE), *Microtrombicula carmenae* (= *Trombicula carmenae*) (PA), *Periglischrus aitkeni* (PA), *Periglischrus ojastii* (MX), *Pseudoschoengastia bulbifera* (PA), *Trichobius brennani* (PA), *Trichobius costalimai* (VE), *Trichobius persimilis* (VE), *Trichobius yunkeri* (PA), *Trombicula soucouyanti* (PA)	Dick et al. (2007), Pacheco et al. (2008b), Sheeler-Gordon and Owen (1999), Wenzel (1976), Wenzel and Tipton (1966)
Sturnira magna (BO,BR,CO, EC,PE)	*Aspidoptera falcata* (PE)	Autino et al. (2011), Pacheco et al. (2008a)
Sturnira mordax (CR,PA)	*Megistopoda* sp. (CR), *Trichobius* sp. (CR)	Timm et al. (1989)
Sturnira oporaphilum (AR,BO, BR,EC,PE)	*Megistopoda proxima* (AR), *Periglischrus ojasti* (BO)	Autino et al. (2009), Barquez and Diaz (2008), Dick et al. (2007)

(continued)

Table 5.1 (continued)

Host (occurrence)	Parasite (occurrence parasite with host)	References
Sturnira sp. (PE)	*Aspidoptera falcate* (PE), *Megistopoda proxima* (PE)	Autino et al. (2011)
Sturnira sp. (BO)	*Megistopoda theodori* (BO), *Periglischrus ojasti* (BO), *Trichobius petersoni* (BO)	Dick et al. (2007)
Sturnira sp. (PA)	*Periglischrus aitkeni* (PA)	Wenzel and Tipton (1966)
Sturnira sp. (VE)	*Trichobius hispidus* (VE)	Wenzel (1976)
Sturnira tildae (BR,PE,VE)	*Aspidoptera falcate* (BR,VE,PE), *Megistopoda proxima* (BR, PE), *Megistopoda* sp. (*proxima* complex) (VE), *Nycterophilia parnelli* (VE), *Strebla guajiro* (VE), *Strebla wiedemanni* (PE), *Trichobius joblingi* (VE), *Trichobius parasparsus* (VE)	Autino et al. (2011), Bertola et al. (2005), Graciolli and Aguiar (2002), Wenzel (1976)
Tonatia bidens (AR,BO,BR, BZ,CR,GT,PY,PA)	*Strebla christinae* (CR), *Strebla galindoi* (CR), *Trichobius* sp. (CR)	Barquez and Diaz (2008), Miller and Tschapka (2009), Timm et al. (1989)
Tonatia saurophila (BR,BZ, PE)	*Strebla galindoi* (BR,PE)	Autino et al. (2011), Dias et al. (2009), Hofstede et al. (2004)
Tonatia sp. (PA)	*Trichobius dugesioides* (PA)	Wenzel and Tipton (1966)
Tonatia sp. (PA)	*Basilia tiptoni* (PA)	Wenzel and Tipton (1966)
Trachops cirrhosis (BO,BR, BZ,CO,CR,EC,GF,GT,GY, HN,MX,NI,PA,PE,SR,SV, VE)	*Ornithodoros brodyi* (PA), *Ornithodoros hasei* (PA), *Paradyschiria curvata* (VE), *Periglischrus tiptoni* (PA), *Periglischrus vargasi* (PA), *Speiseria ambigua* (PA), *Speiseria magnioculus* (VE), *Strebla altmani* (PA), *Strebla carolliae* (PA), *Strebla consocius* (VE), *Strebla diphyllae* (GT), *Strebla kohlsi* (*mirabilis* complex) (PE), *Strebla mirabilis* (BZ,CR,GT, PA,VE), *Trichobius caecus* (VE), *Trichobius dugesii* (PA), *Trichobius dugesioides* (BR,BZ,CR,PA,VE), *Trichobius joblingi* (PA,VE), *Trichobius lionycteridis* (BZ), *Trichobius* sp., closely related to *T. ethophallus* (CR)	Autino et al. (2011), Bärtschi (2000), Dick (2006), Graciolli and Linardi (2002), Hofstede et al. (2004), Miller and Tschapka (2009), Miller et al. (2008), Timm et al. (1989), Wenzel (1976), Wenzel and Tipton (1966)

(continued)

Table 5.1 (continued)

Host (occurrence)	Parasite (occurrence parasite with host)	References
Trinycteris (= *Micronycteris*) *nicefori* (BO,BR,BZ,CO, CR,GF,GT,GY,MX,NI,PA, PE,SR,VE)	*Parastrebla handleyi* (PA), *Strebla alvarezi* (PA), *Strebla obtuse* (PE,VE), *Trichobius dugesii* (VE), *Trichobius joblingi* (PA,VE), *Trichobius keenani* (PA)	Autino et al. (2011), Wenzel and Tipton (1966), Wenzel (1976)
Uroderma bilobatum (BO,BR, BZ,CO,CR,GF,GT,GY,HN, MX,NI,PA,PE,SR,VE)	*Aspidoptera buscki* (VE), *Aspidoptera falcate* (VE), *Basilia constricta* (PE), *Basilia ferrisi* (CO), *Neotrichobius* sp. (*delicatus* complex) (VE), *Neotrichobius stenopterus* (PA), *Ornithodoros hasei* (PA), *Paratrichobius dunni* (BR,BZ,CR,PA,PE,VE), *Periglischrus iheringi* (PA), *Trichobius caecus* (VE), *Trichobius costalimai* (PA), *Trichobius joblingi* (PA,VE), *Trichobius keenani* (PA), *Trichobius longipes* (VE), *Trichobius parasiticus* (VE), *Trichobius parasparsus* (VE), *Trichobius urodermae* (PA,VE)	Autino et al. (2011), Graciolli (2001), Graciolli and Aguiar (2002), Hofstede et al. (2004), Miller and Tschapka (2009), Sampaio et al. (2008f), Wenzel (1976), Wenzel and Tipton (1966)
Uroderma magnirostrum (BO, BR,CO,EC,GT,GY,HN, MX,NI,PA,PE,VE)	*Metelasmus pseudopterus* (VE), *Neotrichobius* sp. (*delicatus* complex) (VE), *Paratrichobius dunni* (VE), *Strebla christinae* (VE), *Strebla wiedemanni* (VE), *Trichobius parasiticus* (VE)	Sampaio et al. (2008b), Wenzel (1976)
Vampyressa bidens (BO,BR, CO,EC,GF,GY,PE,SR,VE)	*Paratrichobius dunni* (PE), *Paratrichobius* sp. (*salvini* complex) (VE)	Autino et al. (2011), Wenzel (1976)
Vampyressa nymphaea (CO, CR,EC,NI,PA)	*Aspidoptera busckii* (PA), *Metelasmus pseudopterus* (PA)	Tavares et al. (2008), Wenzel and Tipton (1966)
Vampyressa pusilla (AR,BR, BZ,CR,PA,VE)	*Neotrichobius delicates* (CR, VE), *Neotrichobius stenopterus* (PA), *Periglischrus iheringi* (PA), *Trombicula dunni* (PA)	Barquez and Diaz (2008), Hofstede et al. (2004), Miller and Tschapka (2009), Wenzel (1976), Wenzel and Tipton (1966)
Vampyressa sp. (PA)	*Paratrichobius dunni* (PA)	Wenzel and Tipton (1966)
Vampyrodes caraccioli (BO, BR,BZ,CO,CR,EC,GF,GT, GY,HN,MX,NI,PA,PE,SR, VE)	*Paratrichobius* sp. (*salvini* complex) (VE), *Periglischrus iheringi* (PA), *Strebla vespertilionis* (PA)	Miller et al. (2008), Wenzel (1976), Wenzel and Tipton (1966)

(continued)

Table 5.1 (continued)

Host (occurrence)	Parasite (occurrence parasite with host)	References
Vampyrum spectrum (BR,BZ, CO,CR,EC,GF,GT,GY,HN, MX,NI,PA,PE,SR,VE)	*Tecomatlana vesperuginis* (= *Hooperella vesperuginis*, = *Trombicula vesperuginis*) (PA), *Trombicula longicalcar* (PA)	Wenzel and Tipton (1966)
Vespertilionidae		
Antrozous pallidus (MX)	*Basilia antrozi* (MX)	Graciolli (2001)
Corynorhinus townsendii (MX)	*Acanthophthirius plecotius* (MX), *Basilia corynorhini* (MX), *Carios yumatensis* (MX), *Cryptonyssus* sp. (MX), *Ichoronyssus* sp. (MX), *Macronyssus cyclaspis* (MX), *Mircotrombicula carmenae* (MX), *Trichobius corynorhini* (MX)	Villegas-Guzman et al. (2005)
Eptesicus brasiliensis (AR,BR, CO,CR,EC,GT,MX,PA,PE, PY,UY,VE)	*Anatrichobius passosi* (BR), *Basilia andersoni* (UY), *Basilia carteri* (BR), *Basilia contricta* (CO), *Basilia currani* (BR), *Basilia mirandariberoi* (BR), *Basilia ortizi* (VE), *Basilia plaumanni* (AR,BR,UY), *Basilia speiseri* (BR), *Basilia wenzeli* (PA,PY)	Autino et al. (2004), Barquez et al. (2008c), Graciolli and Bianconi (2007), Graciolli (2001, 2004), Wenzel and Tipton (1966)
Eptesicus diminutus (AR)	*Basilia neamericana* (AR)	Autino et al. (2009)
Eptesicus furinalis (AR,BO, BR,BZ,CR,GF,GT,GY,HN, MX,NI,PA,PE,PY,SR,UY, VE)	*Basilia andersoni* (UY), *Basilia bequaerti* (BZ,PY), *Basilia carteri* (PY), *Basilia hughscotti* (BR), *Basilia neamericana* (AR), *Basilia ortizi* (BZ,CR,VE), *Basilia plaumanni* (AR,UY), *Basilia wenzeli* (VE), *Hershkovitzia* sp. (BZ)	Autino et al. (2004, 2009), Bärtschi (2000), Barquez et al. (2008c), Graciolli (2001), Graciolli and Aguiar (2002), Miller and Tschapka (2009), Nava et al. (2007)
Eptesicus fuscus (BR,BZ,CO, CR,GT,HN,MX,NI,PA,VE)	*Basilia forcipata* (MX), *Basilia plaumanni* (BR), *Basilia wenzeli* (VE)	Graciolli (2001), Miller et al. (2008)
Eptesicus sp. (VE)	*Basilia ortizi* (VE)	Guimaraes (1977)
Eptesicus sp. (CR,VE)	*Basilia ortizi* (VE), *Basilia* sp. (CR)	Graciolli (2001)
Histiotus laephotis (AR,BO, BR,PE)	*Basilia plaumanni* (AR), *Ornithodoros* sp. (*hasei* complex) (AR)	Autino et al. (2009), Barquez and Diaz (2008), Nava et al. (2007)

(continued)

Table 5.1 (continued)

Host (occurrence)	Parasite (occurrence parasite with host)	References
Histiotus macrotus (AR,BO, BR,CL,PE)	*Basilia plaumanni* (AR), *Basilia* sp. (*velatus?*) (AR)	Autino et al. (2009), Barquez and Diaz (2008), Graciolli et al. (2007)
Histiotus montanus (AR,BO, BR,CL,CO,PY,UY)	*Basilia plaumanni* (PY), *Basilia* sp. *Velatus* (AR)	Gonzalez and Barquez (2008), Graciolli (2001), Graciolli et al. (2007)
Histiotus sp. (UY)	*Basilia plaumanni* (UY)	Graciolli et al. (2007)
Histiotus sp. (CO)	*Basilia wenzeli* (CO)	Graciolli et al. (2007)
Histiotus velatus (AR,BO,BR, CL)	*Basilia andersoni* (BR), *Basilia plaumanni* (BR), *Basilia velatus* (AR,CL), *Megistopoda aranea* (BR), *Ornithodorus* sp. (BR), *Steatonyssus* sp. (BR)	Bertola et al. (2005), Gonzalez and Barquez (2008), Graciolli (2001, 2004), Graciolli et al. (2007), Whitaker and Mumford (1977)
Idionycteris phyllotis (= *Corynorhinus mexicanus*) (MX)	*Carios yumatensis* (MX), *Cryptonyssus* sp. (MX), *Ichoronyssus* sp. (MX), *Macronyssus cyclaspis* (MX), *Microtrombicula carmenae* (MX), *Ornithonyssus pipistrelli* (MX), *Spinturnix americanus* (MX), *Trichobius corynorhini* (MX)	Villegas-Guzman et al. (2005)
Lasiurus borealis (BR,PA,PY)	*Basilia ferruginea* (BR,PA,PY), *Basilia speiseri* (BR)	Graciolli (2001, 2004), Wenzel and Tipton (1966)
Lasiurus castaneus (BR,CR, PA,PY)	*Basilia handleyi* (PA,PY)	Graciolli (2001), Pineda et al. (2008), Wenzel and Tipton (1966)
Lasiurus egregious (BR,CO, PA)	*Strebla carolliae* (PA), *Strebla vespertilionis* (PA)	Sampaio et al. (2008d), Wenzel and Tipton (1966)
Lasiurus sp. (VE)	*Basilia ferrisi* (VE)	Graciolli (2001)
Myotis albescens (AR,BO,BR, CO,CR,EC,GT,GY,HN, MX,NI,PA,PE,PY,SR,UY, VE)	*Amblyomma triste* (UY), *Anatrichobius passosi* (BR), *Anatrichobius scorzai* (AR), *Basilia anceps* (CR), *Basilia andersoni* (BR,UY), *Basilia carteri* (AR,BR,PY,UY), *Basilia constricta* (EC), *Basilia costaricensis* (CR), *Basilia currani* (BR), *Basilia dubia* (PE,VE), *Basilia dunni* (PA,PY,VE), *Basilia ferrisi* (VE), *Basilia manu* (PE), *Basilia mirandariberoi* (BR), *Basilia plaumanni* (BR), *Basilia product* (MX,BR),	Autino et al. (2004, 2009, 2011), Barquez et al. (2008b), Bertola et al. (2005), Dias et al. (2009), Graciolli (2001, 2004), Guimaraes (1977), Maa (1968), Miller and Tschapka (2009), Nava et al. (2007), Timm et al. (1989), Venzal et al. (2008), Wenzel and Tipton (1966)

(continued)

Table 5.1 (continued)

Host (occurrence)	Parasite (occurrence parasite with host)	References
	Basilia sp. (BR), *Basilia typhlops* (VE), *Myodopsylla isidori* (AR,UY), *Myodopsylla wolfssohni* (PE), *Ornithodoros* sp. (*hasei* complex) (AR)	
Myotis californicus (GT,MX)	*Basilia antrozi* (MX)	Arroyo-Cabrales and Perez (2008), Graciolli (2001)
Myotis cf. *albescens* (BR)	*Basilia* sp. (BR)	Graciolli and Linardi (2002)
Myotis cf. *nigricans* (BR)	*Basilia ferrisi* (BR)	Graciolli and Linardi (2002)
Myotis chiloensis (AR,CL,PA)	*Anatrichobius scorzai* (PA), *Basilia currani* (AR), *Basilia silvae* (CL), *Ichoronyssus* sp. (PA), *Joblingia schmidti* (PA), *Macronyssus crosbyi* (= *Ichoronyssus crosbyi*) (PA), *Periglischrus tiptoni* (PA), *Spinturnix americanus* (PA)	Graciolli (2001), Wenzel and Tipton (1966)
Myotis dinellii (AR,BO)	*Myodopsylla isidori* (AR)	Autino et al. (2009), Barquez and Diaz (2008)
Myotis elegans (BZ,CR,GT, HN,MX,NI)	*Basilia anceps* (BZ,CR)	Hofstede et al. (2004), Miller and Rodriguez (2008), Miller and Tschapka (2009), Timm et al. (1989)
Myotis keaysi (AR,BO,BZ,CO, CR,EC,GT,HN,MX,NI,PA, PE,VE)	*Anatrichobius scorzai* (AR,BO, CR,VE), *Basilia anceps* (BZ), *Basilia carteri* (AR), *Myodopsylla isidori* (AR), *Trichobius caecus* (VE)	Autino et al. (2009), Barquez et al. (2008c), Dick et al. (2007), Hofstede et al. (2004), Miller and Tschapka (2009), Wenzel (1976)
Myotis levis (AR,BR,UY)	*Anatrichobius passosi* (BR), *Basilia andersoni* (UY), *Basilia carteri* (AR), *Basilia currani* (AR), *Myodopsylla isidori* (AR,UY)	Autino et al. (2004, 2009), Bertola et al. (2005), Graciolli (2001)
Myotis nesopolus (CO,VE)	*Trichobius costalimai* (VE)	Velazco and Soriano (2008), Wenzel (1976)
Myotis nigricans (AR,BR,BO, CO,CR,EC,GT,HN,MX,NI, PA,PE,PY,UY,VE)	*Anatrichobius scorzai* (BO,CR, PA,VE), *Antricola mexicanus* (PA), *Basilia anceps* (BR,CO, CR,PA,PE,PY,VE), *Basilia andersoni* (BR,UY), *Basilia anomala* (NI), *Basilia carteri* (AR,BO,BR), *Basilia constricta* (EC,PE,VE),	Autino et al. (2009), Barquez et al. (2008c), Bertola et al. (2005), Camilotti et al. (2010), Dick (2006), Dick et al. (2007), Graciolli (2001, 2004), Graciolli and Aguiar (2002), Miller and

(continued)

Table 5.1 (continued)

Host (occurrence)	Parasite (occurrence parasite with host)	References
	Basilia dubia (BR), *Basilia dunni* (PA), *Basilia ferrisi* (CR,CO,GT,GY,PY,VE), *Basilia guimaraesi* (BR), *Basilia hughscotti* (BR), *Basilia juquiensis* (BR), *Basilia lindolphoi* (BR), *Basilia manu* (PE), *Basilia myotis* (PA), *Basilia peruvia* (PE), *Basilia plaumanni* (BR), *Basilia product* (BR), *Basilia rondanii* (GT,HN), *Basilia* sp. (CR), *Basilia* sp. (BO), *Basilia* sp. (BR), *Basilia* sp. (CO), *Basilia speiseri* (BR), *Chiroptonyssus robustipes* (= *Ichoronyssus robustipes*) (PA), *Dermacentor halli* (PA), *Ichoronyssus* sp. (PA), *Joblingia schmidti* (GT,PA), *Megistopoda aranea* (BR), *Megistopoda proxima* (BR), *Metelasmus pseudopterus* (VE), *Myodopsylla isidori* (AR), *Paratrichobius longicrus* (BR), *Rhynchopsyllus pulex* (AR), *Spinturnix americanus* (PA), *Spinturnix* sp. (PA), *Sternopsylla distincta* (PA), *Trichobius parasiticus* (VE), *Trombicula monops* (PA)	Tschapka (2009); own record, Timm et al. (1989), Wenzel (1976), Wenzel and Tipton (1966)
Myotis nigricans or *Myotis chiloensis* (clear determination is not possible, for occurrence see the respective species)	*Anatrichobius scorzai* (PA), *Joblingia schmidti* (PA), *Macronyssus crosbyi* (= *Ichoronyssus crosbyi*) (PA), *Paraspinturnix globosus* (PA), *Spinturnix americanus* (PA), *Sternopsylla distincta* (PA)	Wenzel and Tipton (1966)
Myotis oxyotus (BO,CO,CR, EC,PA,PE,VE)	*Anatrichobius scorzai* (VE)	Aguirre et al. (2008), Wenzel (1976)
Myotis riparius (AR,BO,BR, CO,CR,EC,GF,HN,NI,PA, PE,PY,SR,UY,VE)	*Basilia anceps* (BR,VE), *Basilia andersoni* (UY), *Basilia carteri* (AR), *Basilia ferrisi* (VE), *Basilia hughscotti*	Autino et al. (2004, 2009), Barquez et al. (2008c), Dias et al. (2009), Graciolli (2001, 2004),

(continued)

Table 5.1 (continued)

Host (occurrence)	Parasite (occurrence parasite with host)	References
	(BR), *Basilia juquiensis* (BR), *Basilia manu* (PE), *Basilia ortizi* (VE), *Basilia* sp. (BO), *Basilia* sp. (BR), *Myodopsylla isidori* (UY), *Periglischrus* sp. (BO), *Steatonyssus* sp. (BR)	Graciolli and Aguiar (2002), Graciolli et al. (2007); own record, Whitaker and Mumford (1977)
Myotis ruber (BO,BR,PA,UY)	*Anatrichobius passosi* (BR), *Basilia currani* (BR), *Basilia ruiae* (BR), *Basilia* sp. (BR), *Joblingia* sp. (BR), *Myodopsylla isidori* (BR)	Barquez and Diaz (2008), Bertola et al. (2005), Graciolli (2001, 2004), Graciolli and Bianconi (2007)
Myotis simus (AR,BO,BR,CO, EC,PA,PE,PY,VE)	*Basilia anceps* (PA), *Basilia ferrisi* (VE), *Myodopsylla isidori* (PE,PY,VE), *Spinturnix americanus* (PA)	Barquez and Diaz (2008), Graciolli (2001), Guimaraes (1977), Wenzel and Tipton (1966)
Myotis sp. (PA)	*Anatrichobius scorzai* (PA), *Joblingia schmidti* (PA), *Trombicula soucouyanti* (PA)	Wenzel and Tipton (1966)
Myotis sp. (CR)	*Basilia anceps* (CR)	Miller and Tschapka (2009)
Myotis sp. (BR)	*Basilia andersoni* (BR)	Graciolli (2004)
Myotis sp. (BR)	*Basilia dunni* (BR)	Graciolli and Linardi (2002)
Myotis sp. (BR)	*Basilia currani* (BR)	Graciolli (2004)
Myotis sp. (BR)	*Basilia juquiensis* (BR)	Graciolli (2001)
Myotis sp. (BR)	*Basilia juquiensis* (BR)	Graciolli (2004)
Myotis sp. (BR)	*Basilia lindolphoi* (BR)	Graciolli (2004)
Myotis sp. (GT)	*Trichobius major* (GT)	Dick (2006)
Myotis sp. (MX)	*Basilia traubi* (MX)	Maa (1968)
Myotis sp. (MX)	*Basilia traubi* (MX)	Graciolli (2001)
Myotis sp. (VE)	*Myodopsylla isidori* (VE)	Graciolli (2001)
Myotis thysanodes (MX)	*Myodopsylla isidori* (MX)	Graciolli (2001)
Myotis vivesi (MX)	*Basilia pizonychus* (MX)	Graciolli (2001)
Pipistrellus hesperus (MX)	*Basilia forcipata* (MX)	Graciolli (2001)
Rhogeessa aeneus (BZ,MX)	*Basilia* sp. (BZ)	Arroyo-Cabrales and Ticul Alvarez Castaneda (2008), Hofstede et al. (2004)
Rhogeessa minutilla (CO,VE)	*Trichobius sphaeronotus* (VE)	Soriano and Molinari (2008), Wenzel (1976)
Rhogeessa io (BO,BR,CO, CREC,GF,GY,NI,PA,SR, VE)	*Basilia anceps* (CR)	Miller and Tschapka (2009), Soriano and Tavares (2008)
Rhogeessa tumida (BO,BR,CO, CR,EC,GF,GT,GY,MX,NI, PA,PE,SR,VE)	*Basilia anomala* (VE), *Myodopsylla isidori* (MX,GT, VE), *Spinturnix subacuminatus* (PA)	Graciolli (2001), Guimaraes (1977), Miller et al. (2008), Wenzel and Tipton (1966)

(continued)

Table 5.1 (continued)

Host (occurrence)	Parasite (occurrence parasite with host)	References
Thyropteridae		
Thyroptera discifera (BO,BR, CO,CR,EC,GF,GY,NI,PA, PE,SR,VE)	*Hershkovitzia cabala* (BR), *Hershkovitzia inaequalis* (PE), *Hershkovitzia primitive* (CO,CR), *Myodopsylla isidori* (PE)	Aguirre et al. (2008), Graciolli (2001), Miller and Tschapka (2009)
Bat, no further description		
Alexfainia munozi (PA)		Wenzel and Tipton (1966)
Chiroptonyssus venezolanus (= *Ichoronyssus venezolanus*) (PA)		Wenzel and Tipton (1966)
Euschoengastia desmodus (PA)		Wenzel and Tipton (1966)
Euschoengastia megastyrax (PA)		Wenzel and Tipton (1966)
Eutrombicula goeldii (PA)		Wenzel and Tipton (1966)
Spelaeorynchus latus (PA)		Wenzel and Tipton (1966)
Steatonyssus occidentalis (PA)		Wenzel and Tipton (1966)
Tecomatlana sandovali (PA)		Wenzel and Tipton (1966)
Tecomatlana vesperuginis (= *Hooperella vesperuginis*, = *Trombicula vesperuginis*) (PA)		Wenzel and Tipton (1966)
Trombicula longicalcar (PA)		Wenzel and Tipton (1966)
Trombicula monops (PA)		Wenzel and Tipton (1966)
No host record/host unknown		
Antricola mexicanus (PA)		Wenzel and Tipton (1966)
Basilia sp. (BR)		Graciolli and Linardi (2002)
Basilia bellardii (MX)		Graciolli (2001)
Basilia carteri (PY)		Graciolli (2001)
Basilia constricta (EC,VE)		Graciolli (2001)
Basilia costaricensis (CR)		Graciolli (2001)
Basilia dubia (PE)		Graciolli (2001)
Basilia ferrisi (CR,CO,PY,PE,VE)		Graciolli (2001)
Basilia ferruginea (BR,PY)		Graciolli (2001)
Basilia plaumanni (AR)		Graciolli (2001)
Basilia silvae (CL)		Graciolli (2001)
Basilia speiseri (BR)		Graciolli (2001)
Basilia wenzeli (VE)		Graciolli (2001)
Exastinion clovisi (GT)		Dick (2006)
Hormopsylla kyriophila (PA)		Wenzel and Tipton (1966)
Joblingia schmidti (PA)		Wenzel and Tipton (1966)
Megistopoda aranea (GT)		Dick (2006)
Paratrichobius dunni (GT)		Dick (2006)
Ptilopsylla dunni (PA)		Wenzel and Tipton (1966)
Rhynchopsyllus megastigmata (PA)		Wenzel and Tipton (1966)
Strebla sp. (BR)		Graciolli and Linardi (2002)
Trichobioides perspicillatus (GT)		Dick (2006)
Trichobius caecus (BR)		Graciolli and Linardi (2002)
Trichobius costalimai (GT)		Dick (2006)
Trichobius diphyllae (GT)		Dick (2006)

(continued)

Table 5.1 (continued)

Host (occurrence)	Parasite (occurrence parasite with host)	References
Trichobius dugesii (GT)		Dick (2006)
Trichobius dugesioides (GT)		Dick (2006)
Trichobius longipes (GT)		Dick (2006)
Trichobius parasiticus (GT)		Dick (2006)
Trichobius sparsus (GT)		Dick (2006)
Trichobius uniformis (GT)		Dick (2006)
Trichobius yunkeri (GT)		Dick (2006)
Trichobius sp. (GT)		Dick (2006)
Strebla galindoi (GT)		Dick (2006)
Strebla hertigi (GT)		Dick (2006)
Strebla wiedemanni (GT)		Dick (2006)
Sternopsylla distincta (PA)		Wenzel and Tipton (1966)

Names in parentheses declare synonym name. Country codes: AR = Argentina, BO = Bolivia, BR = Brazil, BZ = Belize, CL = Chile, CR = Costa Rica, CO = Colombia, EC = Ecuador, GT = Guatemala, GF = French Guiana, GY = Guyana, HN = Honduras, MX = Mexico, NI = Nicaragua, PA= Panama, PY = Paraguay, PE = Peru, SR = Suriname, UY = Uruguay, VE = Venezuela

[a]Species split

[b]Species complex (iucnredlist.org)

[c]Tentatively assigned to this species (Dick 2006)

[d]Under taxonomic review

[e]Classification in *Dermanura* not equally accepted

[f]The Central American populations have been split into *Carollia sowelli* (Baker et al. 2002)

publications was equally compiled. However, in such cases the bat may not necessarily have been associated with a parasite. The hosts are listed with their associated parasites according to host family in alphabetical order.

5.3 Contribution to the Bat Ectoparasite Fauna of Bolivia

During the winter months of 2008, 2009 and 2010, we collected 32 ectoparasites from 21 bat individuals of 10 species in the lowlands of Bolivia. The results are shown in Table 5.1. We recorded a total of 16 parasite species. Parasitic insects were represented by nine species of Streblidae and three species of Nycteribiidae. One species of the former and two of the latter family could only be identified to genus level. Representing the Arachnida we found four species of Spinturnicidae, of which two could only be identified to genus level. For species identification, keys of Graciolli (2004) and Miller and Tschapka (2009) were used. All parasites identified to species level were already recorded on the hosts named. For the family Nycteribiidae *Basilia carteri* is shown (Figs. 5.1 and 5.2), and as representatives of the family Streblidae *Megistopoda aranea* (Figs. 5.3 and 5.4) and *Trichobius joblingi* (Figs. 5.5 and 5.6) are shown.

Fig. 5.1 *Basilia carteri*. Dorsal view (scale bar 500 μm). Light micrographs

Fig. 5.2 *Basilia carteri*. Ventral view (scale bar 200 μm). Light micrographs

5.4 Conclusion

In this chapter, the data of the recorded parasites and hosts derived from 20 South and Central American countries. The list shows a large number of bat ectoparasite records from this area. As mentioned before, in case of the two dipteran families Nycteribiidae and Streblidae, we present an extensive share of the worldwide known species. For example, for the last mentioned family we included more than 60 % of the known species. A list of similar extent has not been made for bat ectoparasites before. It is clear that the American leaf-nosed bats (Phyllostomidae) exhibit great numerical superiority not only with respect to host species but also parasite species. An example for the species richness of ectoparasites associated with this family are the 111 parasite species, not including unidentified species, recorded together just for the three bat species *Carollia*

Fig. 5.3 *Megistopoda aranea*. Habitus (scale bar 500 μm). Light micrographs

Fig. 5.4 *Megistopoda aranea*. Ventral view (scale bar 500 μm). Light micrographs

perspicillata, *Desmodus rotundus* and *Sturnira lilium* together. Reasons for this are, amongst others, that they belong to a relatively well-studied bat family and that this family comprises a large part of the bat diversity of the considered areas. As to be

Fig. 5.5 *Trichobius joblingi*. Dorsal view (scale bar 200 μm). Light micrographs

expected, the dipterous families Nycteribiidae and Streblidae form the majority of recorded parasites. The adaptation of both families to the use of bats as hosts is highly specific, but probably not a result of narrow parasite–host coevolution (Graciolli and Carvalho 2012). The authors suggest as one possible reason, that many of their hosts share roosts occasionally at least with closely related species, it is likely that host switching occurs. Much less significant are the mite and tick species. Although they make up the greater part of the parasite fauna of a few hosts, e.g. *Tadarida brasiliensis*, they are normally restricted to a host range that includes only a small number of species. The flea and bug species form the smallest part of the parasite fauna. As temporary parasites, they mostly visit the host only to take a blood meal and thereafter leave it. It has to be mentioned, that for all records the parasite fauna composition differ broadly within the various hosts in the different study areas. Just in a few cases, the same parasite species of one host were recorded simultaneously for different countries. As a result of the different methods of collection applied in the cited studies, probably not all parasite species of the studied bats were recorded. However, the bat species of the Neotropics are still understudied and new distributional records of bats are reported regularly. Consequently, studies on the parasites of bats are incomplete too. The existence of additional parasite species can be assumed.

Fig. 5.6 *Trichobius joblingi*. Ventral view (scale bar 200 μm). Light micrographs

Acknowledgements The present study was financially supported by the research funding programme "LOEWE – Landes-Offensive zur Entwicklung Wissenschaftlich-ökonomischer Exzellenz" of Hesse's Ministry of Higher Education, Research, and the Arts and by the SAW (Senate Competition Committee) grant (SAW-2011-BNI-3) of the Leibniz Association.

References

Aguirre L, Mantilla H, Miller B, Dávalos L (2008) IUCN Red List of threatened species. http://www.iucnredlist.org. Accessed 26 Feb 2013

Aguiar LM, Antonini Y (2011) Descriptive ecology of bat flies (Diptera: Hippoboscoidea) associated with vampire bats (Chiroptera: Phyllostomidae) in the cerrado of central Brazil. Mem Inst Oswaldo Cruz 106(2):170–176. doi:10.1590/S0074-02762011000200009

Arroyo-Cabrales J, Perez S (2008) IUCN Red List of threatened species. http://www.iucnredlist.org. Accessed 27 Feb 2013

Arroyo-Cabrales J, Ticul Alvarez Castaneda S (2008) IUCN Red List of threatened species. http://www.iucnredlist.org. Accessed 27 Feb 2013

Autino AG, Claps GL, González EM (2004) Nuevos registros de insectos (Diptera y siphonaptera) ectoprásitos de murciélagos (Vespertilionidae) del norte de Uruguay. Mastozool Neotrop 11 (1):81–83

Autino AG, Claps GL, Barquez RM, Diaz MM (2011) Ectoparasitic insects (Diptera: Streblidae and Siphonaptera: Ischnopsyllidae) of bats from Iquitos and surrounding areas (Loreto, Peru). Mem Inst Oswaldo Cruz 106(8):917–925. doi:10.1590/S0074-02762011000800004

Autino AG, Claps GL, Barquez RM, Sanchez MS, Diaz MM (2009) New records of bat ectoparasites (Diptera, Hemiptera and Siphonaptera) from northern Argentina. Mem Inst Oswaldo Cruz 106(8):917–925. doi:10.1590/S1519-566X2009000200002

Baker RJ, Solari S, Hoffmann FG (2002) A new Central American species from the *Carollia brevicauda*. Occasional Papers of the Museum of Texas Tech University 217:1–12

Barquez R, Diaz M (2008) IUCN 2012 IUCN Red List of threatened species. http://www.iucnredlist.org. Accessed 21 Feb 2013

Barquez R, Diaz M, Pineda W, Rodriguez B (2008a) IUCN Red List of threatened species. http://www.iucnredlist.org. Accessed 22 Feb 2013

Barquez R, Diaz M, Samudio R, Arroyo-Cabrales J (2008b) IUCN Red List of threatened species. http://www.iucnredlist.org. Accessed 27 Feb 2013

Barquez R, Perez S, Miller B, Diaz M (2008c) IUCN Red List of threatened species. http://www.iucnredlist.org. Accessed 27 Feb 2013

Barquez R, Rodriguez B, Miller B, Diaz M (2008d) IUCN Red List of threatened species. http://www.iucnredlist.org. Accessed 21 Feb 2013

Barquez R, Lim B, Rodriguez B, Miller B, Diaz M (2008e) IUCN Red List of threatened species. http://www.iucnredlist.org. Accessed 21 Feb 2013

Bärtschi D (2000) A study of the Chiroptera of Shipstern Nature Reserve and north-eastern Belize (Central America) together with their ectoparasites (Streblidae, Nycterophiliinea, Acarina) and endoparasties (Cestoda, Nematoda, Trematoda, Acanthocephala). Occasional publication of the International Tropical Conservation Foundation 1–16

Bertola PB, Aires CC, Favorito SE, Graciolli G, Amaku M, Pinto-da-Rocha R (2005) Bat flies (Diptera: Streblidae, Nycteribiidae) parasitic on bats (Mammalia: Chiroptera) at Parque Estadual da Cantareira, São Paulo, Brazil: parasitism rates and host-parasite associations. Mem Inst Oswaldo Cruz 100(1):25–32. doi:10.1590/S0074-02762005000100005

Bruyndonckx N, Dubey S, Ruedi M, Christie P (2009) Molecular cophylogenetic relationships between European bats and their ectoparasitic mites (Acari, Spinturnicidae). Mol Phylogenet Evol 51(2):227–237. doi:10.1016/j.ympev.2009.02.005

Bruyndonckx N, Biollaz F, Dubey S, Goudet J, Christie P (2010) Mites as biological tags of their hosts. Mol Ecol 19(13):2770–2778. doi:10.1111/j.1365-294X.2010.04699.x

Camilotti VL, Graciolli G, Weber MM, Arruda LSJ, Cáceres NC (2010) Bat flies from the deciduous Atlantic forest in southern Brazil: host-parasite relationships and parasitism rates. Acta Parasitol 55(2):194–200. doi:10.2478/s11686-010-0026-2

Claps GL, Autino AG, Barquez RM (2005) Streblidae de murciélagos de Lima: dos citas nuevas para Perú. Rev Sec Entomol Argent 64(1–2):95–98

Dávalos L, Molinari J, Miller J, Rodriguez B (2008a) IUCN Red List of threatened species. http://www.iucnredlist.org. Accessed 21 Feb 2013

Dávalos L, Mantilla H, Medina C, Pineda J, Rodriguez B (2008b) IUCN Red List of threatened species. http://www.iucnredlist.org. Accessed 22 Feb 2013

Dias PA, Caldas dos Santos CL, Rodrigues FS, Rosa LC, Lobato KS, Rebêlo JMM (2009) Espécies de moscas ectoparasitas (Diptera, Hippoboscoidea) de morcegos (Mammalia, Chiroptera) no estado do Maranhão. Rev Bras Entomol 53(1):128–133

Dick CW (2006) The streblid bat flies (Diptera: Streblidae) of Guatemala. In: Canio EB (ed) Biodiversidad de Guatemala, vol 1. University of the Valley of Guatemala, Guatemala City, pp 441–452

Dick CW (2007) High host specificity of obligate ectoparasites. Ecol Entomol 32(5):446–450. doi:10.1111/j.1365-2311.2007.00836.x

Dick CW, Gettinger D (2005) A faunal survey of streblid flies (Diptera: Streblidae) associated with bats in Paraguay. J Parasitol 91(5):1015–1024. doi:10.1645/GE-536R.1

Dick CW, Miller JA (2010) Streblidae (bat flies). In: Brown BV, Wood M, Borkent A, Zumbado M (eds) Manual of Central American Diptera, vol 2. NRC Research Press, Ottawa, ON, pp 1249–1260

Dick CW, Patterson BD (2007) Against all odds: explaining high host specificity in dispersal-prone parasites. Int J Parasitol 37(8–9):871

Dick CW, Gettinger D, Gardner SL (2007) Bolivian ectoparasites: a survey of bats (Mammalia: Chiroptera). Comp Parasitol 74(2):372–377

Dittmar K, Porter ML, Murray S, Whiting MF (2006) Molecular phylogenetic analysis of nycteribiid and streblid bat flies (Diptera: Brachycera, Calyptratae): implications for host associations and phylogeographic origins. Mol Phylogenet Evol 38(1):155–170. doi:10.1016/j.ympev.2005.06.008

Eick GN, Jacobs DS, Matthee CA (2005) A nuclear DNA phylogenetic perspective on the evolution of echolocation and historical biogeography of extant bats (Chiroptera). Mol Biol Evol 22(9):1869–1886

Eriksson A, Graciolli G, Fischer E (2011) Bat flies on phyllostomid hosts in the Cerrado region: component community, prevalence and intensity of parasitism. Mem Inst Oswaldo Cruz 106 (3):274–278. doi:10.1590/S0074-02762011000300004

Esberard CEL, Jesus AC, Motta AG, Bergallo HG, Gettinger D (2005) *Hesperoctenes fumarius* (Hemiptera: Polyctenidae) infesting *Molossus rufus* (Chiroptera: Molossidae) in southeastern Brazil. J Parasitol 91:465–467

Fritz GN (1983) Biology and ecology of bat flies (Diptera: Streblidae) on bats in the genus *Carollia*. J Med Entomol 20(1):1–10

Gonzalez E, Barquez R (2008) IUCN Red List of threatened species. http://www.iucnredlist.org. Accessed 27 Feb 2013

Graciolli G (2001) Distribução geográfica e hospdeiros quirópteros (Mammalia, Chiroptera) de moscas nicteribidas americanas (Diptera, Nycteribiidae). Rev Bras Zool 18(1):307–322

Graciolli G (2004) Nycteribiidae (Diptera, Hippoboscoidea) no Sul do Brasil. Rev Bras Zool 21 (4):971–985. doi:10.1590/S0101-81752004000400035

Graciolli G, Aguiar LS (2002) Ocorrência de moscas ectoparasitas (Diptera, Streblidae e Nycteribiidae) de morcegos (Mammalia, Chiroptera) no Cerrado de Brasília, Distrito Federal, Brasil. Rev Bras Zool 19(1):177–181

Graciolli G, Bianconi GV (2007) Moscas ectoparasitas (Diptera, Streblidae e Nycteribiidae) em morcegos (Mammalia, Chiroptera) em área de Floresta com Araucária no Estado do Paraná, sul do Brasil. Rev Bras Zool 24(1):246–249

Graciolli G, Carvalho CJB (2012) Do fly parasites of bats and their hosts coevolve? Speciation in *Trichobius phyllostomae* group (Diptera, Streblidae) and their hosts (Chiroptera, Phyllostomidae) suggests that they do not. Rev Bras Entomol 56(4):436–450. doi:10.1590/S0085-56262012000400007

Graciolli G, Linardi PM (2002) Some Streblidae and Nycteribiidae (Diptera: Hippoboscoidea) from Maracá Island, Roraima, Brazil. Mem Inst Oswaldo Cruz 97(1):139–141. doi:10.1590/S0074-02762002000100026

Graciolli G, Rui AM (2001) Streblidae (Diptera, Hippoboscoidea) em Morcegos (Chiroptera, Phyllostomidae) no nordeste do Rio Grande do Sul, Brasil. Iheringia Ser Zool 90:85–92. doi:10.1590/S0073-47212001000100009

Graciolli G, Autino AG, Claps GO (2007) Catalogue of American Nycteribiidae (Diptera, Hippoboscoidea). Rev Bras Entomol 51(2):142–159

Graciolli G, Passos FC, Pedro WA, Lim BK (2006) Moscas ectoparasitas (Diptera, Streblidae) de morcegos filostomídeos (Mammalia, Chiroptera) na Estação Ecológica dos Caetetus, São Paulo, Brasil. Rev Bras Zool 23(1):298–299

Guiller A, Deunff J (2010) Spinturnicid mites and bats cophylogeny: comment on Bruyndonckx et al (2009) "Molecular cophylogenetic relationships between european bats and their ectoparasitic mites (Acari, Spinturnicidae)". Mol Phylogenet Evol 57:479–480

Guimaraes LR (1977) Supplementary note on Venezuelan bat flies (Diptera: Nycteribiidae). West N Am Nat 37:221–224

Guzmán-Cornejo C, García-Prieto L, Morales-Malacara JB, Pérez-Ponce De León G (2003) Acarine infracommunities associated with the Mexican Free-Tailed bat, *Tadarida brasiliensis mexicana* (Chiroptera: Molossidae) in arid regions of Mexico. J Med Entomol 40(6):996–999

ter Hofstede HM, Fenton MB (2005) Relationships between roost preferences, ectoparasite density, and grooming behavior of neotropical bats. J Zool Lond 266:333–340. doi:10.1017/S095283690500693X

ter Hofstede HM, Fenton MB, Whitaker JO (2004) Host and host-site specificity of bat flies (Diptera: Streblidae and Nycteribiidae) on neotropical bats (Chiroptera). Can J Zool 82:616–626

Khokhlova IS, Spinu M, Krasnov BR, Degen AA (2004) Immune responses to fleas in two rodent species differing in natural prevalence of infestation and diversity of flea assemblages. Parasitol Res 94(4):304–311

Komeno CA, Linhares AX (1999) Batflies parasitic on some phyllostomid bats in southeastern Brazil: parasitism rates and host-parasite relationships. Mem Inst Oswaldo Cruz 94 (2):151–156. doi:10.1590/S0074-02761999000200004

Krinsky WL (2002) True bugs (Hemiptera). In: Mullen G, Durden L (eds) Medical and veterinary entomology. Elsevier, New York. NY, pp 67–86

Lim BK, Engstrom MD (2004) Mammals of Iwokrama forest. Proc Acad Nat Sci Phila 154:71–108

Linhares AX, Komeno CA (2000) *Trichobius joblingi, Aspidoptera falcata*, and *Megistopoda proxima* (Diptera: Streblidae) parasitic on *Carollia perspicillata* and *Sturnira lillium* (Chiroptera: Phyllostomidae) in southeastern Brazil: sex ratios, seasonality, host site preference, and effect of parasitism on the host. J Parasitol 86(1):167–170. doi:10.1645/00223395(2000)086[0167:TJAFAM]2.0.CO;2

Lourenço CE, Esbérard CEL (2011) Reinfestation of Streblidae ectoparasites (Diptera) in *Carollia perspicillata* (Linnaeus, 1758) (Chiroptera). Rev Bras Parasitol V 20:325–330. doi:10.1590/S1984-29612011000400012

Maa TC (1968) New *Basilia* species from Thailand, Mexico and Brazil (Diptera: Nycteribiidae). Pac Insects 10:25–32

Mans BJ, Louw AI, Neitz AWH (2002) Evolution of hematophagy in ticks: Common origins for blood coagulation and platelet aggregation inhibitors from soft ticks of the genus *Ornithodoros*. Mol Biol Evol 19(10):1695–1705

Mantilla H, Molinari J (2008) IUCN Red List of threatened species. http://www.iucnredlist.org. Accessed 22 Feb 2013

Marshall AG (1982) The ecology of the bat ectoparasite *Eoctenes spasmae* (Hemiptera: Polyctenidae) in Malaysia. Biotropica 14(1):50–55

Miller B, Rodriguez B (2008) IUCN Red List of threatened species. http://www.iucnredlist.org. Accessed 27 Feb 2013

Miller J, Tschapka M (2009) The bat flies of La Selva (Diptera: Nycteribiidae, Streblidae). http://www.biologie.uni-ulm.de/bio3/Batfly/index.html. Last modified 7 May 2009

Miller B, Reid F, Arroyo-Cabrales J, Cuarón AD, de Grammont PC (2008) IUCN Red List of threatened species. http://www.iucnredlist.org. Accessed 26 Feb 2013

Morales-Malacara JB, Colín-Martínez H, García-Estrada CG (2011) A new species of *Eudusbabekia* (Acari: Prostigmata: Myobiidae) from Hart's little fruit bat, *Enchistenes hartii* (Chiroptera: Phyllostomidae), in Southern Mexico. J Med Entomol 48(2):140–145, http://dx.doi.org/10.1603/ME09236

Morales-Malacara JB, Guzmón-Cornejo C, López-Ortega G (2002) A new species of the Genus *Eudusbabekia* (Acari: Prostigmata: Myobiidae) on *Leptonycteris nivalis* (Chiroptera: Phyllostomidae) in Central Mexico. J Med Entomol 39(2):343–349, http://dx.doi.org/10.1603/0022-2585-39.2.343

Munoz P, Fredes F, Raffo E, Gonzales-Acuna D, Munoz L, Cid C (2011) New report of parasite-fauna of the free-tailed bat (*Tadarida brasiliensis*, Geoffroy, 1824) in Chile. Vet Res Commun 35:61–66. doi:10.1007/s11259-010-9449-3

Nava S, Lareschi M, Rebollo C, Benítez Usher C, Beati L, Robbins RG, Durden LA, Mangold AJ, Guglielmone AA (2007) The ticks (Acari: Ixodida: Argasidae, Ixodidae) of Paraguay. Ann Trop Med Parasitol 101(3):255–270. doi:10.1179/136485907X176319

Nikoh N, Sato M, Kondo N, Fukatsu T (2011) Phylogenetic comparison between nycteribiid bat flies and their host bats. Med Entomol Zool 62(3):185–194

Ochoa J, Molinari J (2008) IUCN Red List of threatened species. http://www.iucnredlist.org. Accessed 21 Feb 2013

Ochoa J, Velazco P (2008) IUCN Red List of threatened species. http://www.iucnredlist.org. Accessed 25 Feb 2013

Pacheco V, Aguirre L, Mantilla H (2008a) IUCN Red List of threatened species. http://www.iucnredlist.org. Accessed 26 Feb 2013

Pacheco V, Mantilla H, Pineda J, Rodriguez B (2008b) IUCN Red List of threatened species. http://www.iucnredlist.org. Accessed 26 Feb 2013

Patterson BD, Ballard JW, Wenzel RL (1998) Distributional evidence for cospeciation between neotropical bats and their bat fly ectoparasites. Stud Neotrop Fauna Environ 33:76–84. doi:10.1076/snfe.33.2.76.2152

Patterson BD, Dick CW, Dittmar K (2008) Sex biases in parasitism of neotropical bats by bat flies (Diptera: Streblidae). J Trop Ecol 24:387–396. doi:10.1017/S0266467408005117

Pineda W, Rodriguez B, Samudio R, Pino J (2008) IUCN Red List of threatened species. http://www.iucnredlist.org. Accessed 27 Feb 2013

Poinar G, Brown A (2012) The first fossil streblid bat fly, *Enischnomyia stegosoma* n. g., n. sp. (Diptera: Hippoboscoidea: Streblidae). Syst Parasitol 81(2):79–86. doi:10.1007/s11230-011-9339-2

Rodriguez B, Pineda W (2008) IUCN Red List of threatened species. http://www.iucnredlist.org. Accessed 22 Feb 2013

Rui AM, Graciolli G (2005) Moscas ectoparasitas (Diptera, Streblidae) de morcegos (Chiroptera, Phyllostomidae) no sul do Brasil: associações hospedeiros-parasitos e taxas de infestação. Rev Bras Zool 22(2):438–445. doi:10.1590/S0101-81752005000200021

Salzet M, Capron A, Stefano GB (2000) Molecular crosstalk in host-parasite relationships: schistosome- and leech-host interactions. Parasitol Today 16(12):536–540

Sampaio E, Lim B, Peters S (2008a) IUCN Red List of threatened species. http://www.iucnredlist.org. Accessed 26 Feb 2013

Sampaio E, Lim B, Peters S, Arroyo-Cabrales J (2008b) IUCN Red List of threatened species. http://www.iucnredlist.org. Accessed 22 Feb 2013

Sampaio E, Lim B, Peters S, Rodriguez B (2008c) IUCN Red List of threatened species. http://www.iucnredlist.org. Accessed 26 Feb 2013

Sampaio E, Lim B, Peters S, Samudio R, Pino J (2008d) IUCN Red List of threatened species. http://www.iucnredlist.org. Accessed 27 Feb 2013

Sampaio E, Lim B, Peters S, Velazco P, Molinari J (2008e) IUCN Red List of threatened species. http://www.iucnredlist.org. Accessed 26 Feb 2013

Sampaio E, Lim B, Peters S, Miller B, Cuarón AD, de Grammont PC (2008f). IUCN Red List of threatened species. http://www.iucnredlist.org. Accessed 26 Feb 2013

Sheeler-Gordon LL, Owen RD (1999) Host tracking or resource tracking? The case of *Periglischrus* wing mites (Acarina: Spinturinicidae) of Leaf-Nosed bats (Chiroptera: Phyllostomidae) from Michoacan, Mexico. Acta Zool Mex 76:85–102

Simmons NB (2005) An Eocene big bang for bats. Science 307(5709):527–528. doi:10.1126/science.1108871

Soriano P, Molinari J (2008) IUCN Red List of threatened species. http://www.iucnredlist.org. Accessed 27 Feb 2013

Soriano P, Tavares V (2008) IUCN Red List of threatened species. http://www.iucnredlist.org. Accessed 27 Feb 2013

Tavares V, Muñoz A, Rodriguez B, Arroyo-Cabrales J (2008) IUCN Red List of threatened species. http://www.iucnredlist.org. Accessed 27 Feb 2013

Teeling EC, Springer MS, Madsen O, Bats P, O'Brien J, Murphy WJ (2005) A molecular phylogeny for bats illuminates biogeography and the fossil record. Sci Rep 307:580–584

Timm RM, LaVal RK (1998) A field key to the bats of Costa Rica, vol 22. Occasional Publication Series, University of Kansas, Center of Latin American Studies, pp 1–30

Timm RM, Wilson DE, Clauson BL, La Val RK, Vaughan CS (1989) Mammals of the La Selva-Braulio Carrillo complex, Costa Rica. North American Fauna 75. United States Department of the Interior Fish and Wildlife Service, Washington, DC

Velazco P (2005) Morphological phylogeny of the Bat genus *Platyrrhinus* Saussure, 1860 (Chiroptera: Phyllostomidae) with the description of four new species. Fieldiana Zoology, Chicago, IL

Velazco P, Soriano P (2008) IUCN Red List of threatened species. http://www.iucnredlist.org. Accessed 27 Feb 2013

Velazco P, Muñoz A, Rodriguez B, Pineda W (2008) IUCN Red List of threatened species. http://www.iucnredlist.org. Accessed 26 Feb 2013

Venzal JM, Estrada-Peña A, Castro O, de Souza GG, Félix ML, Nava S, Guigliemone AA (2008) *Amblyomma triste* Koch, 1844 (Acari: Ixodidae): host and seasonality of the vector of *Rickettsia parkeri* in Uruguay. Vet Parasitol 155:104–109. doi:10.1016/j.vetpar.2008.04.017

Villegas-Guzman GA, Lópes-González C, Vargas M (2005) Ectoparasites associated to two species of *Corynorhinus* (Chiroptera: Verspertilionidae) from the Guanaceví Mining region, Durango, Mexico. J Med Entomol 42:125–127

Weirauch C, Gerry AC (2009) Bedbugs. In: Resh VH, Carde RT (eds) Encyclopedia of insects. Elsevier, Oxford, pp 65–66

Wenzel RL (1976) The streblid batflies of Venezuela. Brigham Young University Science Bulletin, Utah

Wenzel RL, Tipton VJ (1966) Appendix. Classified list of hosts and parasites. In: Wenzel RL, Tipton VJ (eds) Ectoparasites of Panama. Field Museum of Chicago, Illinois, pp 797–824

Whitaker JO, Mumford RE (1977) Records of ectoparasites from brazilian mammals. Entomol News 88(9,10):255–258

Whiting MF, Whiting AS, Hastriter MW, Dittmar K (2008) A molecular phylogeny of fleas (Insecta: Siphonaptera): origins and host associations. Cladistics 24:1–31. doi:10.1111/j.10960031.2008.00211.x

Yeates DK, Wiegmann BM (1999) Congruence and controversy: toward a higher-level phylogeny of Diptera. Annu Rev Entomol 44:397–428. doi:10.1146/annurev.ento.44.1.397

Chapter 6
Parasitic Bat Flies (Diptera: Streblidae and Nycteribiidae): Host Specificity and Potential as Vectors

Carl W. Dick and Katharina Dittmar

Abstract Host specificity gauges the degree to which a parasite occurs in association with host species and is among the most fundamental properties of parasite–host associations. The degree of specificity is indicative of myriad properties of the host and parasite and of their ecological and evolutionary relationships. Bat flies are highly specialized bloodfeeding ectoparasites of bats worldwide and were historically viewed as unspecific. In the bat fly—bat system, numerous properties actually or potentially interrupt the linkage of parasite to host and should thus decrease specificity. Such properties of bat flies include a life history strategy requiring females to leave the host, an off-host pupal stage, and high dispersal capability of many species. For hosts, properties include high diversity, mobility, sociality, and multispecies roosting environments. These and other biological and ecological characteristics of bats and flies should together facilitate interspecific host transfers and over time lead to nonspecific host–parasite associations. Despite these properties, large and carefully executed biodiversity surveys of mammals and parasites unequivocally demonstrate the high host specificity of many bat flies, and molecular sequence data promise to demonstrate that many cases of lowered specificity are misunderstood due to unresolved parasite species boundaries. On the other hand, experimental approaches have suggested that host specificity is context dependent and may be lessened in cases of ecological disturbance and in particular when novel host–parasite associations are created. Evolution and maintenance of specificity in bat flies depends in part on the encounter and compatibility properties of bats and on the reproductive potential of the flies on available host species. Moreover, the degree to which parasites have coevolved immunological compatibility with their hosts, thereby diminishing immunological surveillance and response, may also serve to maintain high host

C.W. Dick (✉)
Department of Biology, Western Kentucky University, Bowling Green, KY 42101, USA
e-mail: carl.dick@wku.edu

K. Dittmar
Department of Biological Sciences, University at Buffalo – SUNY, Buffalo, NY 14260, USA

S. Klimpel and H. Mehlhorn (eds.), *Bats (Chiroptera) as Vectors of Diseases and Parasites*, Parasitology Research Monographs 5, DOI 10.1007/978-3-642-39333-4_6,

specificity. Although worldwide bat species on average harbor higher diversity of parasites and pathogens than any other mammalian group, the likelihood of bat flies vectoring disease agents across host species of bats, and particularly to distantly related mammals such as humans, may be relatively small.

Keywords Bat flies • Host specificity • Ectoparasites • Immunological compatibility • Vector • Streblidae • Nycteribiidae

6.1 Introduction

Host specificity is perhaps the most fundamental emergent property of any parasite–host association and is a measure that gauges the degree to which a particular parasite species is limited to, or occurs in strict association with, a particular host species (Poulin 1998a). Parasitic organisms vary in the degree to which they spend time in or on their host, and the degree to which they are found in association with one or more species of host (Marshall 1981). One on extreme, when a parasite species is associated strictly with a single host species, it is understood to be highly specific. Conversely, when a parasite species is associated with a variety of unrelated host taxa, it clearly exhibits low specificity. As a general rule, permanent parasites that spend their entire life cycle on or in a particular host tend to be more specific, whereas temporary parasites that spend little time coupled with their host are less specific (Price 1980). In reality, however, parasitic organisms span a broad spectrum of host specificity. To capture this phenomenon in a more quantitative sense in relation to host taxonomy, Wenzel et al. (1966) introduced the following terms: Monoxenous parasites are restricted to a single host species. Stenoxenous parasites are restricted to a narrow range of closely related hosts (e.g., congeneric hosts). Oligoxenous parasites are restricted to a broader taxonomic range of hosts (e.g., confamilial hosts). Finally, polyxenous ("heteroxenous" sensu Jobling 1949) parasites are not host taxon restricted and occur on a variety of unrelated hosts (Wenzel et al. 1966).

As captured by Wenzel et al. (1966), host specificity is more than a function of the number of host species associations per parasite species but explicitly refers to the host's evolutionary (and taxonomic) associations. It is therefore an attempt to meaningfully incorporate the degree of host relatedness into measures of specificity. Various other measures of specificity (i.e., S_{TD} sensu Poulin and Mouillot 2003) attempt to account for host relatedness. In particular, S_{TD} measures the average taxonomic distinctness of all host species associated with a particular parasite species. While such a measure attempts to capture taxonomic or phylogenetic distinctness, it does not capture ecological distinctness of hosts (Poulin and Mouillot 2003). The degree to which parasites are specific to hosts depends not only on the number of host species or their evolutionary distinctness but also on myriad ecological factors including the parasites' life history strategy, dispersal characteristics, host mobility and social biology, and on dynamics of the nesting or roosting environment of the host (Dick and Patterson 2006; Patterson et al. 2007; Dittmar et al. 2009; Dick and Miller 2010). Finally, there is a strong and intractable relationship between any parasite species' degree of host specificity and the likelihood of that parasite switching to a novel host. Given that many parasites vector pathogens, host-switching is an important means of new host recruitment for parasites and infectious agents alike. This chapter provides an overview of historical and current understanding of the degree of host specificity for a prominent group of bat parasites (Diptera: Nycteribiidae and Streblidae), of the dynamics between bats and these parasites, and finally, the likelihood with which bat flies may be able to move pathogens within and among host species.

6.2 Biology of Bat Flies

6.2.1 Overview

Bat flies (Diptera: Hippoboscoidea) are highly specialized ectoparasites and only associate with bats (Mammalia: Chiroptera) (Figs. 6.1–6.6). They live in the fur and on the wing membranes where they feed on host blood. During the course of their life cycle, however, they necessarily decouple from their host for periods of time, including during female deposition of a prepupae on the roost substrate, and during the nonmotile pupal stage itself. Bat flies are nominally divided into two cosmopolitan families, Streblidae and Nycteribiidae, but an emerging phylogenetic consensus suggests that while bat flies are monophyletic, the families themselves are not natural groups (Dittmar et al. 2006; Petersen et al. 2007). Both families reach the zenith of their diversity within tropical latitudes, are less diverse in the subtropics, and rather impoverished in temperate regions. However, this latitudinal richness gradient appears more pronounced in the Western Hemisphere. Nycteribiids (ca. 275 species) are more speciose in the Eastern Hemisphere, whereas the streblids (ca. 230 species) are richer in the Western Hemisphere (Dick and Patterson 2006).

6.2.2 Life History

Relatively little is known regarding life history and reproductive biology of bat flies, and what is known is based on a very limited number of studies involving only a few species. Nonetheless, some generalizations likely apply to all bat flies. Hippoboscoid Diptera (including bat flies) reproduce via viviparous puparity (Hagan 1951). Eggs are fertilized inside the female, and the three larval stages develop within the female, likely nourished by intrauterine "milk" glands. When internal development is complete, female flies leave the host bat, seek and find a

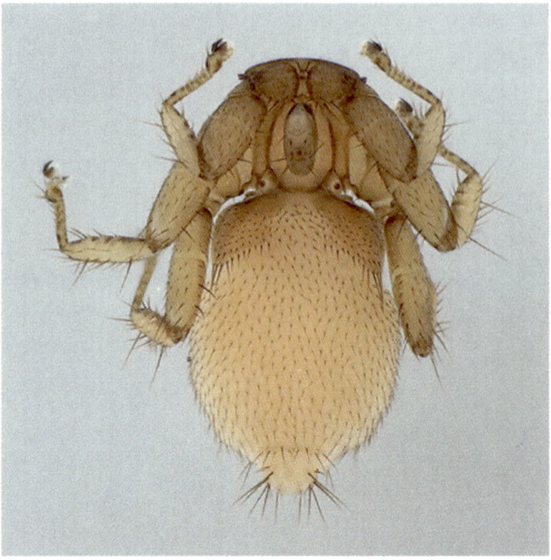

Fig. 6.1 *Archinycteribia octophthalma* Theodor (Nycteribiidae: Archinycteribiinae) from *Penthetor lucasi*, female, dorsal view (Sarawak)

Fig. 6.2 *Cyclopodia inclita* Falcoz (Nycteribiidae: Cyclopodiinae) from *Pteropus tonganus*, male, dorsal view (Fiji)

suitable deposition site (typically, within the confines of the bat's roosting environment), and deposit a single, terminal (3rd instar) larva on this substrate. Once deposited, the larva ("prepupa") immediately forms a puparium. Following a pupal stage ca. 3–4 weeks in duration (Ching and Marshall 1968), an unfed (teneral) adult fly emerges and must locate and colonize a host. Ching and Marshall (1968) reported prepupal deposition directly on the host's body. But in these atypical cases, the pupae were distorted in shape and most failed to develop into an adult fly or were removed by host grooming.

The most thorough treatment of the life history of a bat fly species is Overal's (1980) study of the life cycle of *Megistopoda aranea* (Streblidae) in Panama. In this species, prepupae were deposited in the roost, but usually in close proximity of bats (*Megistopoda* spp. are incapable of flight). Following an average of 23 days, the adult fly emerged from the puparium and located a host for a blood meal before

Fig. 6.3 *Nycterophilia* n. sp. (Streblidae: Nycterophiliinae) from *Brachyphylla cavernarum*, female, lateral view (St. John, US Virgin Islands)

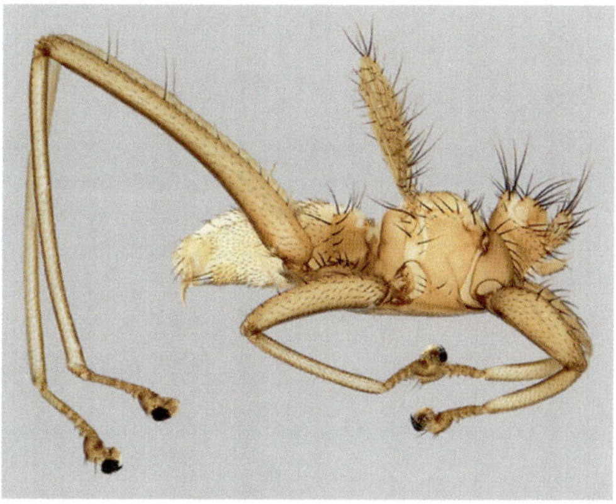

Fig. 6.4 *Megistopoda aranea* (Coquillett) (Streblidae: Trichobiinae) from *Artibeus jamaicensis*, male, lateral view (Panama)

mating. Postpartum females were observed mating within minutes after depositing a prepupa. The time interval between successive depositions of prepupae was about 10 days. Another detailed study was conducted by Fritz (1983), on the Costa Rican species *Trichobius joblingi*, *Speiseria ambigua*, and *Strebla guajiro* (all Streblidae). These species also deposited prepupae on roost substrates away from the host, with the pupal stage duration of 17, 20, and 19 days for these species, respectively. Recently, Dittmar et al. (2009, 2011) confirmed prepupal deposition away from the roost for other *Trichobius* spp. (Mexico, Puerto Rico), as well as *Nycterophilia* spp. (Puerto Rico). A few studies also have been conducted on temperate (North American) streblid species. Generally, life histories of temperate species mirrored those in tropical regions. Unlike tropical bats, however, temperate bats hibernate during winter months, and bat flies remained physically and reproductively active on hibernating bats whose body temperatures ranged from 2 to 6 °C (Ross 1960; Reisen et al. 1976; Caire and Hornuff 1986).

Fig. 6.5 *Aspidoptera phyllostomatis* (Perty) (Streblidae: Trichobiinae) from *Artibeus planirostris*, female, dorsal view (Ecuador)

Fig. 6.6 *Megastrebla gigantea* (Speiser) (Streblidae: Brachytarsininae) from unknown host, female, dorsal view (Philippines)

The life cycle of *Basilia hispida* (Nycteribiidae) has been described in some detail by Marshall (1970). In this species, flies reached sexual maturity 5–6 days after emergence from the puparium. Like that observed in the streblid *Megistopoda aranea*, male *B. hispida* often not only copulated with postpartum females immediately following prepupal deposition but also were observed copulating with newly emerged and unfed (teneral) females. A single copulation was sufficient to produce several offspring, suggesting that female *B. hispida* may store sperm. At intervals of ca. 9 days, mature females deposited prepupae on roost substrate, which were depressed onto the substrate by use of their abdomen. Prepupal deposition occurred between 0900 and 1800 h, apparently stimulated in part by temperatures elevated by the presence of bats in the roost. Development of pupae depended upon whether bats were present or absent. When bats were present, teneral adults emerged as soon as 25 days after deposition. When bats were absent or their presence sporadic, pupal development was nearly twice as long (up to 46 days). Newly emerged adults began walking about (all nycteribiids are wingless) until encountering a host. Upon colonization of a host, teneral flies began feeding within 20 min. The total life span of *B. hispida* averaged 136 and 195 days for males and females, respectively, with 5 days prepartum, 9 days in the larval stages, 25 days in the pupal stage, and 97 days (males) or 156 days (females) in the adult stage (Marshall 1970). When experimentally removed from hosts, *B. hispida* died within 5–25 h; mature females were less resistant after removal from the host than were teneral flies.

The breeding biology of another nycteribiid, *Eucampsipoda sundaica*, was described by Ching and Marshall (1968). Most features of the life cycle resembled that of *B. hispida*, but during mating, males attached to females for up to ½ h, during which time the female remained fully ambulatory. The interval between successive prepupal depositions was three times faster in *E. sundaica* than *B. hispida* (3 versus 9 days, respectively) (Ching and Marshall 1968). Peterson and Wenzel (1987) suggested that the life cycle of bat flies was rather uniform across species. In general, the life history strategy of bat flies appears to reflect their obligate association with bats, where vulnerable immature stages generally remain coupled with the host inside the female fly, and completion of the life cycle, constrained by the relatively vulnerable pupal stage, requires a somewhat stable roosting environment away from the risks of host grooming (Patterson et al. 2007).

6.2.3 Host as Habitat

One prevailing or defining characteristic of parasitism is that the parasite is to some degree dependent upon their host. As such, the host itself is often considered to be the "habitat" for the parasite. The host's body itself provides a broad array of environmental stimuli and resources to the parasite and plays a major role in shaping the life history, ecology, and behavior of the parasite. Moreover, the habitat affects dynamics of parasite populations, assemblages, or community-level interactions among multiple parasite species (Patterson et al. 2008a, b; Tello et al. 2008; Dick and Miller 2010).

Certain parasites are highly dependent on habitat variables such as host body size and correlated characters. This is known as "Harrison's Rule" (Eichler 1963) and states that size of parasites and hosts are positively correlated. This rule is evident in highly habitat-specific parasites such as bird wing lice (Phthiraptera), where lice transferred to larger or smaller hosts experience significant reduction in fitness (Bush and Clayton 2006). Although bat flies are not as host microhabitat specific as bird wing lice, they do exhibit a degree of general habitat specificity, living on either the wing and tail membranes, or in the fur, of the bats that they infest (Dick and Patterson 2006). In these general habitats, they appear to feed in a similar manner but tend to conform to three general morphological forms (one adapted to patagia and two adapted to fur) that are related to different evasion strategies to counter-effect host grooming efforts. Although such specificity to particular host macrohabitats may help maintain some degree of specificity, across host taxa these broad categories (e.g., patagia and haired regions) of macrohabitat should not prevent cross-host switching of flies that specialize on particular macrohabitats. The nature and specificity of parasitic associations is strongly influenced by the ecology and behavior of both host and parasite (Brooks and McLennan 1993; Marshall 1981; Poulin 1998a).

6.2.4 Roost as Habitat

Variation in parasitic life history traits and host specificity may result from the capacity of the parasite to perceive and respond to a much larger set of environmental variables than those presented strictly by the host itself (Heeb et al. 2000). In addition to the behavior and ecology of both associates, environmental features shape the dynamics of parasite fitness. The greater environment of parasites has dimensions and linkages that extend from its immediate environment such as the host itself to the surrounding ecosystem in which the host lives (Pilosof et al. 2012). Many parasitological studies have focused on the "host as habitat" and have largely ignored the greater nesting or roosting environment of hosts. While a plethora of studies document and explore the former, the latter is still poorly understood and documented. For example, studies of fleas have highlighted instances where parasitism varied with soils and other habitat variables (Krasnov et al. 1997, 1998) and where such relationships are more strongly associated with host species than with habitat variables (Krasnov et al. 2006). Flea species, however, vary tremendously in the permanence of their coupling with hosts (Medvedev and Krasnov 2006; Whiting et al. 2008).

The dimensionality of the off-host environment varies greatly with the degree to which host and parasite remain in direct physical contact throughout their life cycles. For example, all phases of the life cycle of bird and mammal chewing lice take place strictly on a host individual. Lice have drastically reduced visual systems, are relatively immobile off the host, and typically exhibit high fidelity to individual host species (Clayton et al. 2004). These characteristics often correspond with observations of parallel clade formation, or cospeciation (Hafner and Page 1995; Hafner et al. 1998; Reed et al. 2007; Light and Reed 2009). Ixodid ticks, conversely, spend up to 90 % of their lives off-host and most species parasitize a broad range of vertebrate hosts (Klompen et al. 1996; Poulin 1998b). When such a parasite's life cycle interrupts the linkage with a host, its potential for ecological proliferation and coevolution is limited. Bat flies, although obligate ectoparasites, are holometabolous and must metamorphose in order to complete their development. Given that pupation takes place off the host, in the bat's roost, and lasts a few weeks, there is a necessary decoupling of parasite and host. This means that bat flies spend possibly up to a third of their adult life span off the host and in the immediate roosting environment, mainly as pupae (Dick and Patterson 2006). In order to be specific to a particular host species, newly emergent flies must locate not just any bat but a bat of a particular species.

Roosting habits of bat species vary greatly in relation to bat fly development. Although colony size, composition, occupancy, and interbat spacing may all be expected to influence parasitism, of particular importance is the structure of roosts themselves. The types of roosts that bats choose span remarkable ecological amplitude. Different species seek shelter in a variety of roosts, including leaf tents, tree foliage, bark, and cavities, termite nests, rock crevices, houses, mines, culverts, and caves (Kunz and Lumsden 2003). Considering this variety of roosting structures, they vary in durability from very short lived to permanent. For example, while foliage-type roosts offer abundant and ubiquitous roosting locations, they tend to be ephemeral and bats spend only a few days or weeks in a given roost (Timm 1987). At the other extreme roosts such as caves or mines, while often uncommon and patchily distributed, represent somewhat permanent roosting spaces that can be utilized by many successive generations of bats (Kunz 1982). Roosts also vary in the amount of exposure and protection they offer roosting bats, in a manner that should affect the roost fidelity of bats: foliage and leaf tents are relatively exposed and open, while caves mines are enclosed and protected.

The roosting biology of bats thus influences the quality and quantity of parasitism by bat flies (Lewis 1995; Marshall 1981; ter Hofstede and Fenton 2005; Reckardt and Kerth 2006; Patterson et al. 2007), which in turn influences the likelihood of vectoring pathogens. Because bat flies pupate in the roost itself, the duration of bat residency at roosts determines to a great degree the likelihood that newly emerged flies are reassociated with the original host population of their parents. Although many factors affect roost-site fidelity in bats (Lewis 1995), durability must play an important role. Both ecological and coevolutionary associations of bats and bat flies thus depend on the fly's ability to place pupae in a location conducive to emergent offspring quickly and reliably reassociating with individuals of the host species (Dittmar et al. 2011). Roost fidelity of bats also varies according to the durability of its roost and the shelter it affords from predation and adverse weather (Kunz 1982; Kunz and Lumsden 2003). In this context, it has been shown that the number of bat fly species parasitizing a bat, the proportion of host individuals parasitized, and the number of parasites per infested host all increase with the permanence and protection of bat roosts (Patterson et al. 2007). However, when investigating the mobility of bat flies in relation to the aforementioned variables, it was found that dispersal ability as measured by the presence or absence of functional wings did not affect evolutionary associates, or the prevalence and intensity of parasitization by bat flies (Patterson et al. 2007). Although such variables have not been assessed in the light of host specificity among bat flies, we expect this measure to be unrelated also to dispersal ability of the parasites.

6.2.5 Historical View of Host Specificity

The degree to which nycteribiid and streblid bat flies are host specific has long been debated (Jobling 1949; Theodor 1957; Wenzel et al. 1966; Marshall 1981; Poulin 1998a; ter Hofstede et al. 2004). Early studies (Jobling 1949; Theodor 1957) insisted that bat flies are not particularly host specific. At the time, only 9 of the 36 known species of New World streblid flies were thought of as strictly monoxenous and even then, only one species was documented sufficiently to confirm its level of specificity (Jobling 1949). Thus, monoxeny of the majority of those nine species was doubtful. In an apparent nod to the importance of roosting dynamics, lack of specificity was explained due to the fact that many species of bats often roost together, giving flies ample opportunity to parasitize multiple host species (Jobling 1949). Close and consistent proximity of host bats should increase the likelihood of exchange of parasites among different bat taxa, where bat flies would effectively utilize multiple taxa as one large host metapopulation. Jobling (1949) noted that host preference was distinct among some fly species, and he forwarded the concept of the "principal host" to describe the observed preference of certain flies for certain host species. However, it was claimed that some fly species had "so completely adapted" to different species and genera of bats that it was impossible to ascertain a principal host (Jobling 1949).

It has become increasingly clear that historical treatments of host specificity among bat flies were hampered by several factors, including (1) limited collection records, (2) poor identifications (and limited taxonomic understanding) of both parasites and hosts, (3) lack of voucher collections of both parasites and hosts, and (4) poorly controlled collection protocols to minimize interhost contamination. It has been argued that the single most important requirement for determining host specificity is the systematic assembly of large, geographically representative collections (Wenzel et al. 1966). Indeed, early assessments of bat fly specificity primarily were based on haphazardly accumulated museum collections, gathered over long periods of time and by numerous workers whose foci often were not on issues regarding host specificity (Marshall 1976). During the initial phase of one large survey (Panama; Wenzel et al. 1966), field workers routinely placed bat specimens of multiple species into the same receptacle, and only later would these hosts be sampled for ectoparasites. Moreover, surveys routinely employed workers not properly trained in collection of ectoparasites (R. L. Wenzel, personal communication).

Nonetheless, the early studies set the foundation for our understanding of host specificity. Based on a sample of 12,000 Panamanian streblid flies, Wenzel et al. (1966) found that about 55 % of the 66 streblid species were associated with a single host species (monoxenous sensu stricto), and that an additional 15 % were monoxenous (sensu *lato*), known rarely to "stray" onto ecologically associated hosts such as roost associates. Of the remaining 30 %, about half were oligoxenous and the remaining 15 % were polyxenous. A decade later, a very large and systematic collection effort was undertaken in Venezuela (Smithsonian Venezuelan Project), resulting in ca. 36,000 streblid specimens. Wenzel (1976) treated 116 Venezuelan bat fly species and assessed host specificity. Seventy-five percent of those species were strictly monoxenous, while the remaining 25 % were steno-, oligo-, or polyxenous. In Malaysia, Marshall (1980) reported that 72 % and 64 % of streblids and nycteribiids, respectively, were recorded from a single host, with the remaining fly species restricted to sets of congeneric species (stenoxenous). It was during this time and based on the results of very large collections that workers on bat flies began to refute claims of low specificity forwarded previously by Jobling (1949) and Theodor (1957).

6.3 Contemporary View of Host Specificity

6.3.1 Field Surveys and Natural History Collections

In general and as collection data have grown in both quality and quantity, the trend has been one of an increasing recognition of host specificity for bat flies (Marshall 1981; Hutson 1984; ter Hofstede et al. 2004; Dick 2007; Dick and Patterson 2007). In the past couple of decades, surveys of mammals and ectoparasites have been designed and executed specifically to assess host specificity (among other things) and with an eye to reducing potential for cross-host contamination of parasite samples (Presley 2004; Dick and Gettinger 2005). Such survey protocols mandated strict isolation of each host individual and that cleanliness steps are taken to enhance our ability to accurately discern patterns in host–parasite specificity. Nevertheless, our knowledge of specificity among ectoparasites largely depends on mammal–ectoparasite surveys that are broad in scope and varied in objective. Given the nature of such surveys, literature based on them often contains dubious host–parasite associations (e.g., Wenzel et al. 1966; Wenzel 1976; Autino et al. 1999; Komeno and Linhares 1999; Graciolli and de Carvalho 2001; Dick and Gettinger 2005).

Regarding sampling and host specificity, increased acknowledgment of the reality and likelihood of contamination invoked the use of terms such as "nonprimary" or "transitory" hosts (e.g., Wenzel et al. 1966; Wenzel 1976). Often, such nonprimary or transitory associations were assumed and the cause of such associations not pursued further. They were explained as: (1) rare but natural host transfers, (2) disturbance (accidental) transfers made while the host or parasite was being sampled, or (3) contamination from parasite sampling events (Wenzel et al. 1966). Usually no attempt was made to discriminate among the possible causes of such associations. Moreover, other workers sometimes ignored these issues and listed indiscriminately any parasite species reported from a host species (e.g., Guerrero 1997). Even when host–parasite associations are deemed nonprimary, such assessments may be based on raw percentages of flies on hosts (e.g., Wenzel 1976). Statistical assessments of host specificity of ectoparasites remain rare. While host-specificity indices have been proposed (Miles et al. 1957; Stark and Kinney 1969; Rohde 1980), they all can be problematic and unreliable, particularly in cases where parasite species are monoxenous and others stenoxenous and oligoxenous (Poulin 1998a). Understanding the nature of host–parasite associations depends upon the accuracy of host specificity appraisals. The degree to which parasite species are specific greatly informs studies of host–parasite ecology, evolution, and cospeciation, and as such, data regarding host specificity and interhost contamination of parasites must be addressed quantitatively. Quantitative assessments of host specificity would allow the use of statistical tests to determine the likelihood that observed associations between host and parasite are ecological phenomena or whether they could occur by chance. Moreover, insights into historical patterns and processes, most notably phylogenetically framed ones, are obscured when reliable and accurate estimates of host specificity under natural conditions are unavailable (Brooks and McLennan 1993).

Two general scenarios exist to explain the association of a given parasite species and a given host species. First, the parasite may have intrinsically come to be associated with that host (e.g., by colonization of that particular host after emergence from a puparium), or by vertical transfer in a cophyletic sense (e.g., the parasite is associated with a particular host because its mother is associated with that host). Alternatively, it may be in association with that host species due to human influence of some kind (i.e., it was contaminated onto a host via inadequate collection methods, or made the host switch under its own power but while it or its host was under duress upon collection). Discriminating between these two alternative explanations (natural vs. human mediated) is the critical first step to understanding host–parasite specificity. Furthermore, determination of whether nonprimary host associations are natural or a result of human intervention can identify the cases when parasites are inclined to switch hosts naturally (see Sect. 6.3.4) and direct future studies of the ecology and evolution of host–parasite associations.

In a recent survey specially designed to eliminate cross-host contamination of parasites, Dick and Gettinger (2005) determined that >99 % of nearly 2,500 fly individuals sampled in Paraguay were associated with primary host species. Using information on primary and nonprimary host captures per day, and assuming flies could not become contaminated on days they were not collected, all but one of the

15 mismatches were shown to have resulted from human error upon sampling contamination (Dick 2007). In other words, naturally caused host-switching appeared to be rare indeed. Accounting for these cases of human error, the level of host specificity observed for the Paraguay collection was higher than that in any other study published to date on bat flies. In that study, 27 of the 31 bat fly species (87 %) were strictly monoxenous. Had the nonprimary associations observed during the Paraguay study been mistakenly reported as primary associations, the observed degree of monoxeny would have been only 71 %.

In addition to sampling contamination, another cause of nonspecific associations is disturbance transfer. Disturbance transfers are difficult to control, avoid, and assess. The vast majority of bat captures involves the use of harp traps, mist nets, and handheld loop nets (Kunz et al. 2009). Typically, nets are deployed along trails, streams, and other flyways frequented by bats (Kunz and Kurta 1988). When a bat is captured, it experiences some degree of physiological stress and will often struggle to escape. Moreover, bats of different species are often captured simultaneously (particularly in harp traps) and will be in close proximity for a period of time. Bat flies are sensitive to the stress experienced by the bat, and often become agitated and leave the host during this time (R. L. Wenzel, personal communication; personal observation). Particularly among winged bat fly species (~75 % of New World streblid species are capable of flight; Whitaker et al. 2009), there is an increased likelihood of disturbance transfers from one bat species to another. Finally, when field workers handle and remove bats from nets or traps, the bats become further stressed and the ectoparasites further disturbed. Flies commonly move onto the hands of bat handlers, or fly toward the handler's light source. As it is with highly host-specific gopher lice (Hafner et al. 2003), dispersal capability was long held to be an important factor in the degree of host specificity, in bat flies with flightless species being more host specific, and winged species being less specific (Jobling 1949). In general, flightless species only able to crawl from host to host should be more host limited than volant species that may easily move from host to host over great distances. However, nycteribiids (which are all wingless) appear to be no more host specific than streblids (of which ~25 % of the species are incapable of flight) (Marshall 1980). Marshall (1976) regarded both families of bat flies as being "host specific" with species overwhelmingly monoxenous but occasionally stenoxenous. Recently, ter Hofstede et al. (2004) suggested that mobility has no effect on the degree of host specificity among Belizean bat flies. All said, these dynamics serve to further obscure collections-based assessments of host specificity because they serve to increase the likelihood of unnatural transfers of what otherwise may be host-specific parasites. Frequent or even constant monitoring of mist nets and harp traps may reduce the likelihood of disturbance transfers but cannot eliminate them entirely.

Recent collections from Ecuador (unpublished data) provide some additional evidence that host specificity among bat flies is high, particularly when assessed under controlled conditions. Based on information collected from >2,000 bats of 77 species and >6,000 bat flies of 97 species, criteria for bat fly sample size (20 or more per fly species) resulted in 42 fly species of sufficient number. For the multiyear and multi-investigator collections countrywide, 17 fly species were strictly monoxenous, 22 were stenoxenous, and three were polyxenous. In other words, ~93 % of well-sampled Ecuadorian bat flies parasitize a single host species or no more than a few closely related host species. To illustrate the importance of standardization and strict control of collection protocol, the Ecuadorian data was divided into two sets. One set comprised all data known from collecting efforts in Ecuador, which included all samples from multiple collection events over multiple years in Ecuador, some of which are old and employed little if any control over sampling. These data were collected by numerous individuals, who lacked parasitological training, and who were less likely to be aware of the necessity for sampling protocols to ensure true primary host associations were determined. The second set is a subset of the total data that included only samples from one large but recent collection (Sowell Expedition) where all of the sampling protocols were known, several appropriate measures were taken to avoid sampling error, and all samples were collected by only two people (CWD and a student under direct supervision). This subset of the data includes more than ½ of the total dataset. Using the same sampling parameters described above, 30 bat fly species met the criteria, and a picture of even greater host specificity emerges. It was found that out of 30 bat fly species; 17 were strictly monoxenous, while 13 were stenoxenous. None were observed to be oligo- or polyxenous. These results strongly indicate that Ecuadorian bat fly species are quite specific to individual species or genera of bats and are consistent with emerging consensus of high host specificity among bat flies.

Clearly, well-executed collections of ectoparasites that minimize interhost contamination, and the statistical evaluation of nonprimary host associations, both contribute to the increased understanding of high host specificity among bat flies. Typically, realistically high levels of specificity are observed only when (1) large enough samples of hosts and parasites are available, (2) sampling protocols strictly control for interhost sampling contamination, and (3) instances of nonprimary host associations are evaluated statistically.

6.3.2 Experimental and Molecular Assessments

Although survey and collection-based assessments of specificity reveal high host specificity among bat flies, molecular and experimental assessments have the potential to obscure and complicate such an understanding. In one of the few published experimental studies of streblid bat flies and their choice of hosts, cafeteria-choice arenas were established to allow flies to choose among alternative host options (Dick and Dick 2006). Results of this study revealed strong preference of flies for host bats that varied in quantity and quality of previous infestation. Similar results were obtained in an unpublished study by Dittmar (in 2009) for the nycteribiid *Leptocyclopodia* sp. in a host choice experiment between the bat species *Cynopterus brachyotis* and *Pteronotus jagori* in the Philippines. In a more natural setting, specificity of bat flies may also depend in part on whether alternative host species live allopatrically or sympatrically. In eight caves in Sri Lanka, Seneviratne et al. (2009) investigated specificity of bat flies parasitizing *Rousettus leschenaulti*, *Rhinolophus rouxi*, and *Hipposideros speoris*. In two caves, the bat species were sympatric, but in six other caves (2, 2, and 2, respectively) the bat species roosted alone. Generally, local monoxeny was reported for seven of eight species of nycteribiid, and for five of six species of streblid. However, certain bat fly species were five times more likely to show strong host preferences when host species were roosting together, whereas less specificity was evinced on host species roosting alone (Seneviratne et al. 2009). These preferences were evident in both nycteribiids and streblids and may reflect strong selection for fly fitness (higher fitness on specific hosts, but lower fitness on less-preferred hosts when only that host is available). Such selection may be driven by competition among parasite species or by differential nutritive properties of the host blood. Laboratory experiments with parasite transfers, however, highlight the distinct possibility of parasite establishment on novel hosts in the absence of the various ecological and evolutionary dynamics typically preventing encounter and establishment (Perlman and Jaenike 2003; Bush and Clayton 2006). This trend is important in light of bat flies as pathogen vectors, and the likelihood at which new emerging infectious diseases may spillover to novel hosts; to the degree to which environmental disturbance and/or novel host–host contact occurs, the potential with which parasites may spill over and establish on novel hosts increases.

While experimental studies, particularly those done in the laboratory, suggest that host specificity is not evolutionarily "fixed", in some cases molecular scrutiny attests to even higher specificity than is seen through field survey data alone (Poulin and Morand 2005). In fact, the existence of morphologically cryptic species diversity is being increasingly discovered in what were previously understood to be less-than-monoxenous parasites (Whiteman et al. 2006; Smith et al. 2006; Sect. 6.3.4). Such assessment has not been done to date with bat flies, but the existence of widespread cryptic species speaks to the potential for currently understood steno-, oligo-, or polyxenous bat fly species to actually comprise multiple monoxenous species that are currently indistinguishable morphologically. Such a situation, if it comes to be known, would greatly increase the operational host specificity of bat flies and further decrease their likelihood to transmit pathogens across host species boundaries.

6.3.3 Maintenance of Specificity

Many factors are at play with respect to the cause of host specificity, which includes both the evolution and maintenance of specificity in parasites. Various proximate factors have been identified, such as nutritive and immunological compatibility of blood biochemistry, sensory cues that allow a parasite to recognize a specific host (e.g., olfactory and thermal cues emanating from the host and the greater environment), or the morphological adaptations of parasites that facilitate physical establishment and maintenance on the host's body landscape. Ultimate causes of host specificity include the ability to find and utilize food and mates, evolutionary constraints upon vertically transmitted parasite taxa that have cospeciated with their hosts, or any other suite of factors that increase the reproductive fitness of the parasite over evolutionary time.

The evolution of specificity in parasites centers on the dynamics of parasite dispersal, host proximity availing parasite establishment, and adaptation of the parasites themselves (Timms and Read 1999). As a lineage of mammals, bats harbor an incredible array of arthropod parasite diversity (e.g., Acari [mites and ticks], Diptera, Dermaptera, Hemiptera, and Siphonaptera) and particularly for parasite diversity of higher taxonomic rank (e.g., order and family) appear to harbor more parasite diversity than any other group of mammals (Wenzel and Tipton 1966). As mentioned above, the roosting environment of bats plays a profound role in shaping dynamics of parasitism. Throughout the evolutionary history of bats and their potential parasites, surely the utilization of relatively stable and long-lived roosting sites has facilitated many independent transitions to parasitism among arthropods associated with bats. Long-lived and dependable roost sites, coupled with copious and steady supplies of organic matter in the form of guano and dead carcasses, as well as presence of the bats themselves, has facilitated the evolution of parasitism among roost-associated arthropod groups (e.g., Siphonaptera). Such roosting dynamics would have facilitated arthropod dispersal to bats and increased encounter rates, and the duration of these roosting sites would have allowed strong selection to shape the myriad adaptations among bat parasites that we observe today.

As a general rule, host-specific parasites typically are limited in their dispersal capacity (e.g., lice on solitary fossorial mammals; Hafner et al. 2003) or are morphologically, behaviorally, or physiologically adapted to their host to such a degree that survival on a novel host may be severely diminished (Tompkins and Clayton 1999; Bush and Clayton 2006). Thus, the evolution and maintenance of host specificity among bat flies is probably fueled by selective compromise. On the one hand, additional potential host species represent additional resources for population establishment, and exploiting them should increase both abundance and overall fitness of the parasite (Poulin 1998a). On the other hand, broad exploitation strategies would expose such a parasite to a larger array of physical and immunological challenges as well as potential competitors for niche space, in turn diminishing such a parasite's ability to fine tune its exploitation to the ecology and behavior of any one primary host species. Given the dispersal capabilities of bat flies and the social structure of their bat hosts, specificity in bat flies is likely adaptive rather than being produced by intrinsic dispersal limitations (Dick and Patterson 2007).

The evolution of parasite life cycles and scenarios for why a particular parasite species might be present or absent on a particular host species was outlined by Combes (1991). This framework later became known as the filter concept (Poulin 1998b). The paradigm has direct application to the evolution and maintenance of host specificity in bat flies. Combes (1991) described two filters that independently constrain potential parasites from establishing on a host species. First, the encounter filter serves to exclude potential hosts the parasite cannot locate, encounter, or colonize for behavioral or ecological reasons. For bat parasites, for example, reliable and long-lived roosting environments would serve to make the encounter filter porous by facilitating likelihood of encounter and colonization of bats by potential parasites. The encounter filter embodies parasite dispersal limitations described by Timms and Read (1999). The second filter, the compatibility filter, represents another threshold in the evolution of parasitism, by excluding all host individuals on which the parasite cannot survive for morphological, physiological, or immunological reasons (Combes 1991). The compatibility filter embodies parasite adaptive limitations (Timms and Read 1999). Together, the encounter and compatibility filters would each have to be passed before parasite populations were established, and before any degree of host specificity can be reached (Combes 1991). The filters constrain the pool of potential parasite taxa.

As such, the filter concept is largely host centric (e.g., the host must be encountered, and the host must be compatible) and may not account for characteristics intrinsic to the parasite or potential parasite. In any parasite, one factor of crucial importance to evolutionary fitness is the efficacy with which they can encounter and successfully mate with other members of the same species. To incorporate this aspect of the evolution and maintenance of host specificity, the reproductive filter was proposed (Dick and Patterson 2006). This filter serves to exclude all host individuals on which the parasite cannot find mates, or upon which they are subject to decreased reproductive fitness. On the basis of mate availability and reproductive potential, high host specificity of bat flies should reinforce selection for continued specificity. If potential mates are absent from nonprimary hosts, the reproductive potential of colonizing flies would be zero, despite having passed the encounter and compatibility filters with respect to nonprimary hosts. Flies that transferred to nonprimary hosts are more likely to perish without reproducing, unless they move back onto a primary host. If indeed bat flies, or certain lineages of bat flies, are cospeciated with their hosts (Patterson et al. 1998), host speciation is likely to have been allopatric and would have forced reproductive isolation of fly populations isolated on new host populations. So while potential cospeciation may in some cases cause reproductive isolation, the reproductive filter could serve to maintain such isolation. Indeed, even transient movement of flies onto nonprimary hosts appears in general to be exceedingly rare in ecological time (Dick 2007). In other words, flies may be constrained to specificity because their primary host is the only place where they consistently and successfully find their mates. This would suggest that a reproductive filter may act as an additional threshold to the successful establishment and maintenance of parasite–host specificity.

It is possible that bat flies may also have developed immunocompatibility with specific host species. This has been demonstrated in other bloodfeeding parasites (e.g., fleas), where more generalist parasites infested host birds with weakened immune systems, while more host-specific flea species were able to feed on certain hosts regardless of the status of their immune systems (Møller et al. 2004). This points to the potential existence of shared antigenic epitopes between host and parasite, where each associate may use the same or similar immune-signaling molecules to avoid aggressive immunosurveillance. In bat flies, it has long been held that irritation from the mechanical damage of bites and potentially from introduced saliva may provoke grooming responses, which represent a source of ectoparasite mortality (Marshall 1981). Risk of mortality from host grooming may be higher for relatively large bat flies (ca. 1–5 mm), given that the hosts of many are insectivorous in their feeding habits. For permanent, host-specific parasites, selection should work to reduce irritation and thus grooming response from parasite feeding bouts. Although such studies have not been undertaken with bat flies, the presence of nested immunological compatibility in diverging host clades may go far to explain the well-known phylogenetic pattern of host specificity and the maintenance thereof.

6.3.4 Breakdown of Specificity

Although the general trend among streblid and nycteribiid flies appears to be one of high host specificity, there appear to be certain cases where bat flies depart from the overall pattern. As discussed above, field survey and natural history collections data reveal repeated (although limited) instances of steno-, oligo-, and polyxenous bat fly species. For example, 3 of 31 streblid fly species surveyed in Paraguay were less specific than strictly monoxenous (Dick and Gettinger 2005). Looking more closely at these species, they include *Trichobius jubatus*, *Aspidoptera phyllostomatis*, and *Megistopoda aranea*. Each of these species appears to be a special case. First, *T. jubatus* are rarely encountered flies in the *Trichobius dunni* species group (sensu Wenzel 1976) that parasitize several genera and species of New World molossid bat. *Trichobius dunni* group species are the only bat flies of new world molossid bats and interestingly, molossids are the only New World bats that host another group of bloodfeeding parasites, the bat bugs (Hemiptera: Polyctenidae; Whitaker et al. 2009). In Paraguay, *T. dunni* parasitized *Molossus molossus*, *M. rufus*, *Eumops patagonicus*, *E. glaucinus*, and *Molossops temminckii*. Prevalence rates on these bat species were typically low, ranging from <1 % to 15.8 %. On at least three of these host species, there is no evidence that records could possibly be due to sampling contamination. However, whether *T. jubatus* actually comprises several cryptic species, or whether the species truly is oligoxenous remains unknown at this time. But among the ca. 210 species of western hemisphere bat flies, *T. jubatus* is among the least specific species, parasitizing at least three genera of molossid bats. Its associations certainly are exceptional. As for the species *Aspidoptera phyllostomatis* and *Megistopoda aranea*, these stenoxenous species tend to co-occur on the same host species, which are the phyllostomid bats *Artibeus fimbriatus* and *A. planirostris*. A closely related species, *Megistopoda proxima*, occurs on numerous species of *Sturnira* and is known to be a complex of closely related species (Wenzel 1976) that are separable morphometrically, but species boundaries are not diagnosable (CW Dick, unpublished data). Sequence data from quickly evolving markers should serve to illuminate these units of biodiversity, and it is predicted that these stenoxenous species (like others, see Poulin and Keeney 2007) will in fact turn out to be closely related monoxenous species.

Another relatively rare situation where host specificity appears to truly break down occurs in "hot roost caves" of Puerto Rico and elsewhere in the American Antilles (Rodríguez-Durán 1998). These particular cave roosts are unique and characterized by temperatures between 28 and 40 °C and relative humidity exceeding 90 %. The high temperatures are understood to result from the metabolic activity of numerous bats and a cave topology that allows entrapment of heat (Rodríguez-Durán 1998). Within these caves, bats are parasitized by species of the streblid genus *Nycterophilia*, which are associated with bats belonging to the families Mormoopidae, Natalidae, and Phyllostomidae. *Nycterophilia* flies appear to exclusively parasitize obligate cave-roosting bats and often parasitize bats inhabiting "hot roosts." Recently, fly ecology and host associations of flies and their endosymbiotic *Gammaproteobacteria* in Puerto Rican hot caves were evaluated in the context of host specificity and roost specificity (Morse et al. 2012a). Likelihood-based ancestral character mapping provided evidence that symbionts were originally associated with bat flies that facultatively used both hot cave main chambers and cooler cave foyers. Later, the *Nycterophilia* and their symbionts evolved to an association exclusively with hot caves and bat species therein. In addition, this transition appeared to be accompanied by flies switching from being more host and less habitat specific to flies being more habitat specific and less host specific (Morse et al. 2012a). If these transitions represent the true evolutionary history of these bats, flies, and endosymbionts, it suggests selective pressure on the environmental tolerance of particular life history stages of bat flies. This situation may be a rare case of an evolutionary "retrogression" away from high host specificity among bat flies.

6.3.5 Potential as Vectors

As obligate bloodfeeding parasites, bat flies would appear to be excellent candidates for vectors of zoonotic agents. Moreover, bats are numerically abundant in the environment, evolutionarily diverse, and geographically widespread and as such are reservoirs to a stunning variety of viruses and other potential pathogens (Calisher et al. 2006). Even compared to the more speciose Rodentia, bat species on average harbor more pathogens than those of any other mammal group (Luis et al. 2013). The biology, ecology, and behavior of bats (e.g., dietary breadth, colonial sociality, mobility and migration, roost choice amplitude, etc.) make them ideal reservoir hosts for viruses and other potential disease agents. That bats are subject to higher rates and diversity of pathogen infestation than other mammal groups may owe to their high rates of sympatry, which should encourage cross-reservoir spillover of pathogens among species. Bat flies and other bloodfeeding arthropods are often suspected as playing a role in such spillover. To anyone who is experienced with handling bats, bat flies are perhaps the most consistently noticeable of the bloodfeeding arthropod parasites. They are typically large (1–5 mm), active, and stand out against bat fur and patagia. But for their size, bloodfeeding habits, and often high rates prevalence and intensity, it is currently unclear if bat flies are well positioned to moving pathogens across host species, let alone in transference to humans. Though there appears to be ample evidence of lineage sorting among bat and bat fly associations over evolutionary time, the rate at which flies naturally switch hosts in ecological time appears to be miniscule, although this is not a well-researched topic. If this is true, however, this would greatly diminish the potential for flies to serve as interhost species vectors of pathogens.

However, some of the characteristics owing to why bats have so many pathogens in the first place may speak to the likelihood of increased host switching under bat duress and disturbance. In cases of habitat or roost disturbance, bats that are otherwise ecologically separated may increase the frequency and intensity of contact with novel species. This disturbance in turn may cause an increase of accidental or transitory associations of bat flies with nonprimary or nonnatural hosts. In contrast to otherwise normal ecological conditions, such dynamics may in fact increase likelihood of host switching and hence, of movement of pathogens to novel host species.

Bats often host complex ectoparasite assemblages, and there is anecdotal evidence that some of these parasites harbor and transmit bacteria and other pathogens. For example, the bat tick *Carios kelleyi* is known to vector a number of pathogenic bacteria (*Rickettsia*, *Borrelia*, and *Bartonella*) (Loftis et al. 2005); cimicid ectoparasites of bats are vectors of *Trypanosoma* (Paterson and Woo 1984); and nycteribiids are known to transmit *Polychromophilus*, a protozoan parasite, to bats (Gardner and Molyneux 1988).

Given host specificity of bat flies, what does seem abundantly probable is that bat flies may be implicated in intrahost reservoir transference and maintenance of pathogens. For example, species of both nycteribiids and streblids are known to infest *Hypsignathus monstrosus, Epomops franqueti*, and *Myonycteris torquata*, three species of Old World fruit bats (Pteropodidae) that were reported to harbor Ebola virus (Leroy et al. 2005). Within a host population, bat flies readily move from bat to bat, feeding on multiple individuals during the course of days and weeks (Overal 1980). If flies can actually uptake viral pathogens and those pathogens remain viable within bat flies, it is probable that even host-specific flies transfer such viruses among host bats within a population.

The burgeoning literature on arthropod endosymbionts points to a multitude of bacteria and other agents in bat flies that may be mutualistic or pathogenic (Trowbridge et al. 2006; Nováková et al. 2009; Hosokawa et al. 2012; Morse et al. 2012a, b; Billeter et al. 2012). Specifically, recent research has identified *Bartonella* spp. in bat flies and bats. Bartonellae have been implicated in zoonoses in humans (Harms and Dehio 2012). The role of *Bartonella* spp. for bat flies and bats is unclear at the time, but their function as symbionts or pathogens (for bat flies) and pathogens (for bats) are being considered. There is also strong evidence that among bat fly symbionts, at least some are capable of being transmitted vertically from mother to offspring, rather than being passed laterally by uptake from the environment at large (Hosokawa et al. 2012; Morse et al. 2013). Moreover, phylogenies of some of the endosymbionts (e.g., Candidatus *Aschnera chinzeii*) are highly congruent with that of their nycteribiid bat fly hosts (Hosokawa et al. 2012) lending credence to the hypothesis that the associates are cospeciated, and that transmission of the bacteria is indeed vertical rather than horizontal. Although these results do not rule out the possibility of horizontal transmission of pathogenic agents by bat flies, they do suggest tightly coevolved linkages between bat flies, and associated symbionts and pathogens.

References

Autino AG, Claps GL, Barquez RM (1999) Insectos ectoparasitos de murciélagos de las Yungas de la Argentina. Acta Zool Mex 78:119–169

Billeter SA, Hayman DTS, Peel AJ, Baker K, Wood JLN, Cunningham A, Suu-Ire R, Dittmar K, Kosoy MY (2012) *Bartonella* species in bat flies (Diptera: Nycteribiidae) from western Africa. Parasitology 139:324–329

Brooks DR, McLennan DA (1993) Parascript: parasites and the language of evolution. Smithsonian Institution, Washington, DC, 429 pp

Bush SE, Clayton DH (2006) The role of body size in host specificity: reciprocal transfer experiments with feather lice. Evolution 60:2158–2167

Caire W, Hornuff L (1986) Overwintering population dynamics of the bat fly *Trichobius major* (Diptera: Streblidae). Southwest Nat 31:126–129

Calisher CH, Childs JE, Field HE, Holmes KV, Shountz T (2006) Bats: important reservoir hosts of emerging viruses. Clin Microbiol Rev 19:531–545

Ching LM, Marshall AG (1968) The breeding biology of the bat-fly *Eucampsipoda sundaicum* Theodor, 1955 (Diptera: Nycteribiidae). Malayan Nat J 21:171–180

Clayton DH, Bush SE, Johnson KP (2004) Ecology of congruence: past meets present. Syst Biol 53:165–173

Combes C (1991) Evolution of parasite life cycles. In: Toft CA, Aeschlimann A, Bolis L (eds) Parasite-host associations: coexistence or conflict? Oxford University Press, Oxford, pp 62–82

Dick CW (2007) High host specificity of obligate ectoparasites. Ecol Entomol 32:446–450

Dick CW, Dick SC (2006) Effects of prior infestation on host choice of bat flies (Diptera: Streblidae). J Med Entomol 43:433–436

Dick CW, Gettinger D (2005) A faunal survey of streblid bat flies (Diptera: Streblidae) associated with bats in Paraguay. J Parasitol 91:1015–1024

Dick CW, Miller JA (2010) Streblidae. In: Brown BV, Borkent A, Cumming JM, Wood DM, Woodley NE, Zumbado M (eds) Manual of central American diptera, vol II. National Research Council, Ottawa, pp 1249–1260

Dick CW, Patterson BD (2006) Bat flies: obligate ectoparasites of bats. In: Morand S, Krasnov B, Poulin R (eds) Micromammals and macroparasites: from evolutionary ecology to management. Springer, Berlin, pp 179–194

Dick CW, Patterson BD (2007) Against all odds: explaining high host specificity in dispersal-prone parasites. Int J Parasitol 37:871–876

Dittmar K, Porter ML, Murray S, Whiting MF (2006) Molecular phylogenetic analysis of nycteribiid and streblid bat flies (Diptera: Brachycera, Calyptratae): implications for host associations and phylogeographic origins. Mol Phylogenet Evol 38:155–170

Dittmar K, Dick CW, Patterson BD, Whiting MF, Gruwell ME (2009) Pupal deposition and ecology of bat flies (Diptera: Streblidae): *Trichobius* sp. (caecus group) in a Mexican cave habitat. J Parasitol 95:308–314

Dittmar K, Morse S, Gruwell M, Mayberry J, DiBlasi E (2011) Spatial and temporal complexities of reproductive behavior and sex ratios: a case from parasitic insects. PLoS One 6(5):e19438. doi:10.1371/journal.pone.0019438

Eichler WD (1963) Arthropoda. Insecta. Phpthiraptera. I. Mallophaga. In: Bronns HG (Ed) Klassen und Ordnungen des Tierreichs. III. Insecta. 7b. Pthiraptera. Akademische Verlagsanstalt Geest und Portig, Leipzig, pp 158–179

Fritz GN (1983) Biology and ecology of the bat flies (Diptera: Streblidae) on bats of the genus *Carollia*. J Med Entomol 20:1–10

Gardner RA, Molyneux DH (1988) *Polychromophilus murinus* – a malarial parasite of bats – life history and ultrastructural studies. Parasitology 96:591–605

Graciolli G, de Carvalho CJB (2001) Moscas ectoparásitas (Diptera, Hippoboscoidea) de morcegos (Mammalia, Chiroptera) do Estado do Paraná. II. Streblidae. Chave pictórica para gêneros e espécies. Rev Bras Zool 18:907–960

Guerrero R (1997) Catálogo de los Streblidae (Diptera: Pupipara) parásitos de murciélagos (Mammalia: Chiroptera) del Nuevo Mundo. VII. Lista de especies, hospedadores y paises. Acta Biol Venez 19:9–24

Hafner MS, Page RDM (1995) Molecular phylogenies and host parasite coevolution: gophers and lice as a model system. Philos Trans R Soc Lond B Biol Sci 349:77–83

Hafner MS, Demastes JW, Hafner DJ, Spradling TA, Sudman PD, Nadler SA (1998) Age and movement of a hybrid zone: implications for dispersal distance in pocket gophers and their chewing lice. Evolution 52:278–282

Hafner MS, Demastes JW, Spradling TA, Reed DL (2003) Cophylogeny between pocket gophers and chewing lice. In: Page RDM (ed) Tangled trees: phylogeny, cospeciation and coevolution. University of Chicago Press, Chicago, pp 195–220

Hagan HR (1951) Embryology of the viviparous insects. Ronald, New York

Harms A, Dehio C (2012) Intruders below the radar: molecular pathogenesis of *Bartonella* spp. Clin Microbiol Rev 25:42–78

Heeb P, Kölliker M, Richner H (2000) Bird-ectoparasite interactions, nest humidity, and ectoparasite community structure. Ecology 81:958–968

Hosokawa T, Nikoh N, Koga R, Satô M, Tanahashi M, Meng X-Y, Fukatsu T (2012) Reductive genome evolution, host–symbiont co-speciation and uterine transmission of endosymbiotic bacteria in bat flies. ISME J 6:577–587

Hutson AM (1984) Keds, flat-flies and bat-flies (diptera, hippoboscidae and nycteribiidae). Handbook for the identification of British insects. Royal Entomological Society of London, London, 40 pp

Jobling B (1949) Host-parasite relationship between the American Streblidae and the bats, with a new key to the American genera and a record of the Streblidae from Trinidad, British West Indies (Diptera). Parasitology 39:315–329

Klompen JSH, Black WC, Keirans JE, Oliver JH (1996) Evolution of ticks. Annu Rev Entomol 41:141–161

Komeno CA, Linhares AX (1999) Batflies parasitic on some phyllostomid bats in southeastern Brazil: parasitism rates and host-parasite relationships. Mem Inst Oswaldo Cruz 94:151–156

Krasnov BR, Shenbrot GI, Medvedev SG, Vatschenok VS, Khokhlova IS (1997) Host–habitat relations as an important determinant of spatial distribution of flea assemblages (Siphonaptera) on rodents in the Negev Desert. Parasitology 114:159–173

Krasnov BR, Shenbrot GI, Khokhlova IS, Medvedev SG, Vatschenok VS (1998) Habitat dependence of a parasite-host relationship: flea (Siphonaptera) assemblages in two gerbil species of the Negev Desert. J Med Entomol 35:303–313

Krasnov BR, Stanka M, Miklisova D, Morand S (2006) Habitat variation in species composition of flea assemblages on small mammals in central Europe. Ecol Res 21:460–469

Kunz TH (1982) Roosting ecology. In: Kunz TH (ed) Ecology of bats. Plenum, New York, pp 1–55

Kunz TH, Kurta A (1988) Capture methods and holding devices. In: Kunz TH (ed) Ecological and behavioral methods for the study of bats. Smithsonian Institution, Washington, DC, pp 1–29

Kunz TH, Lumsden LF (2003) Ecology of cavity and foliage roosting bats. In: Kunz TH, Fenton MB (eds) Bat ecology. University of Chicago Press, Chicago, pp 3–89

Kunz TH, Hodgkison R, Weise CD (2009) Methods of capturing and handling bats. In: Kunz TH, Parsons S (eds) Ecological and behavioral methods for the study of bats. Johns Hopkins University Press, Baltimore, pp 3–35

Leroy EM, Kumulungui B, Pourrut X, Rouquet P, Hassanin A, Yaba P, Delicat A, Paweska JT, Gonzalez JP, Swanepoel R (2005) Fruit bats as reservoirs of Ebola virus. Nature 438:575–576

Lewis SE (1995) Roost fidelity of bats: a review. J Mammal 76:481–496

Light JE, Reed DL (2009) Multigene analysis of phylogenetic relationships and divergence times of primate sucking lice (Phthiraptera: Anoplura). Mol Phylogenet Evol 50:376–390

Loftis AD, Gill JS, Schriefer ME, Levin ML, Eremeeva MW, Gilchrist MJR, Dasch GA (2005) Detection of *Rickettsia, Borrelia*, and *Bartonella* in *Carios kelleyi* (Acari: Argasidae). J Med Entomol 42:473–480

Luis AD, Hayman DT, O'Shea TJ, Cryan PM, Gilbert AT, Pulliam JR, Mills JN, Timonin ME, Willis CK, Cunningham AA, Fooks AR, Rupprecht CE, Wood JL, Webb CT (2013) A comparison of bats and rodents as reservoirs of zoonotic viruses: are bats special? Proc R Soc B 280:20122753. doi:10.1098/rspb.2012.2753

Marshall AG (1970) The life cycle of *Basilia hispida* Theodor 1957 (Diptera: Nycteribiidae) in Malaysia. Parasitology 61:1–18

Marshall AG (1976) Host-specificity amongst arthropods ectoparasitic upon mammals and birds in the New Hebrides. Ecol Entomol 1:189–199

Marshall AG (1980) The comparative ecology of insects ectoparasitic upon bats in west Malaysia. In: Wilson DE, Gardner AL (eds) Proceedings of the 5th international bat research conference. Texas Tech, Lubbock, TX, pp 135–142

Marshall AG (1981) The ecology of ectoparasitic insects. Academic, London, p 459

Medvedev SG, Krasnov BR (2006) Fleas: permanent satellites of small mammals. In: Morand S, Krasnov BR, Poulin R (eds) Micromammals and macroparasites: from evolutionary ecology to management. Springer, Tokyo, pp 161–177

Miles VIA, Kinney R, Stark HE (1957) Flea-host relationships of associated *Rattus* and native wild rodents in the San Francisco Bay area of California, with special reference to plague. Am J Trop Med Hyg 6:752–760

Møller AP, Marti M, Vivaldi N, Soler JJ (2004) Parasitism, host immune defence and dispersal. J Evol Biol 17:603–612

Morse S, Dick CW, Patterson BD, Dittmar K (2012a) Some like it hot: evolution and ecology of novel endosymbionts in bat flies of cave-roosting bats (Hippoboscoidea, Nycterophiliinae, *Nycterophilia*). Appl Environ Microbiol 78:8639–8649

Morse S, Olival K, Kosoy M, Billeter S, Patterson BD, Dick CW, Dittmar K (2012b) Global distribution and genetic diversity of *Bartonella* in bat flies (Hippoboscoidea, Streblidae, Nycteribiidae). Infect Genet Evol 12:1717–1723

Morse SF, Bush S, Patterson BD, Dick CW, Gruwell M, Dittmar K (2013) Evolution, multiple acquisition, and localization of endosymbionts in bat flies (Hippoboscoidea, Streblidae, Nycteribiidae). Appl Environ Microbiol 79:2952–2961. doi:10.1128/AEM.03814-12

Nováková E, Hypsa V, Moran NA (2009) *Arsenophonus*, an emerging clade of intracellular symbionts with a broad host distribution. BMC Microbiol 9:143. doi:10.1186/1471-2180-9-143

Overal WL (1980) Host-relations of the batfly *Megistopoda aranea* (Diptera: Streblidae) in Panama. Univ Kansas Sci Bull 52:1–20

Paterson WB, Woo PTK (1984) The development of the culture and blood-stream forms of 3 *Trypanosoma* (*Schizotrypanum*) spp. (Protista, Zoomastigophorea) from bats in *Cimex lectularius* (Hemiptera, Cimicidae). Can J Zool 62:1581–1587

Patterson BD, Ballard JWO, Wenzel RL (1998) Distributional evidence for cospeciation between Neotropical bats and their bat fly ectoparasites. Stud Neotrop Fauna Environ 33:76–84
Patterson BD, Dick CW, Dittmar K (2007) Roosting habits of bats affect their parasitism by bat flies (Diptera: Streblidae). J Trop Ecol 23:177–189
Patterson BD, Dick CW, Dittmar K (2008a) Parasitism by bat flies (Diptera: Streblidae) on Neotropical bats: effects of host body size, distribution and abundance. Parasitol Res 103:1091–1100
Patterson BD, Dick CW, Dittmar K (2008b) Sex biases in parasitism of neotropical bats by bat flies (Diptera: Streblidae). J Trop Ecol 24:387–396
Perlman SJ, Jaenike J (2003) Infection success in novel hosts: an experimental and phylogenetic study of *Drosophila*-parasitic nematodes. Evolution 57:544–557
Petersen FT, Meier R, Kutty SN, Wiegmann BM (2007) The phylogeny and evolution of host choice in the Hippoboscoidea (Diptera) as reconstructed using four molecular markers. Mol Phylogenet Evol 45:111–122
Peterson BV, Wenzel RL (1987) Nycteribiidae. In: McAlpine JF, Peterson BV, Shewell GE, Teskey HJ, Vockeroth JR, Wood DM (eds) Manual of nearctic Diptera. Research Branch, Agriculture Canada, Ottawa, pp 1283–1291
Pilosof S, Dick CW, Corine C, Patterson BD, Krasnov B (2012) Effects of anthropogenic disturbance and climate on patterns of bat fly parasitism on bats. PLoS One 7:1–7. doi:10.1371/journal.pone.0041487
Poulin R (1998a) Evolutionary ecology of parasites: from individuals to communities. Chapman and Hall, London, 212 pp
Poulin R (1998b) Host and environmental correlates of body size in ticks (Acari: Argasidae and Ixodidae). Can J Zool 76:925–930
Poulin R, Keeney DB (2007) Host specificity under molecular and experimental scrutiny. Trends Parasitol 24:24–28
Poulin R, Morand S (2005) Parasite Biodiversity. Smithsonian Institution, Washington, 224 pp
Poulin R, Mouillot D (2003) Parasite specialization from a phylogenetic perspective: a new index of host specificity. Parasitology 126:473–480
Presley SJ (2004) Ectoparasitic assemblages of Paraguayan bats: ecological and evolutionary perspectives. Ph.D. Dissertation, Texas Tech University, Lubbock, TX, 326 pp
Price PW (1980) Evolutionary biology of parasites. Monogr Popul Biol 15:1–237
Reckardt K, Kerth G (2006) The reproductive success of the parasitic bat fly *Basilia nana* (Diptera: Nycteribiidae) is affected by the low roost fidelity of its host, the Bechstein's bat (*Myotis bechsteinii*). Parasitol Res 98:237–243
Reed DL, Light JE, Allen JM, Kirchman JJ (2007) Pair of lice lost or parasites regained: the evolutionary history of anthropoid primate lice. BMC Biol 5:1–11
Reisen WK, Kennedy ML, Reisen NT (1976) Winter ecology of ectoparasites collected from hibernating *Myotis velifer* (Allen) in southwestern Oklahoma (Chiroptera: Vespertilionidae). J Parasitol 62:628–635
Rodríguez-Durán A (1998) Nonrandom aggregations and distribution of cave-dwelling bats in Puerto Rico. J Mammal 79:141–146
Rohde K (1980) Host specificity indices of parasites and their application. Experientia 36:1369–1371
Ross A (1960) Notes on *Trichobius corynorhini* on hibernating bats (Diptera: Streblidae). Wasmann J Biol 18:271–272
Seneviratne SS, Fernando HC, Udagama-Randeniya PV (2009) Host specificity in bat ectoparasites: a natural experiment. Int J Parasitol 39:995–1002
Smith MA, Woodley NE, Janzen DH, Hallwachs W, Hebert PDN (2006) DNA barcodes reveal cryptic host-specificity within the presumed polyphagous members of a genus of parasitoid flies (Diptera: Tachinidae). Proc Natl Acad Sci USA 103:3657–3662
Stark HE, Kinney AR (1969) Abundance of rodents and fleas as related to plague in Lava Beds National Monument, California. Pan-Pac Entomol 38:249–251

Tello JS, Stevens RD, Dick CW (2008) Patterns of species co-occurrence and density compensation: a test for interspecific competition in bat ectoparasite infracommunities. Oikos 117:693–702

ter Hofstede HM, Fenton MB (2005) Relationships between roost preferences, ectoparasite density, and grooming behaviour of Neotropical bats. J Zool 266:333–340

ter Hofstede HM, Fenton MB, Whitaker JO Jr (2004) Host and host-site specificity of bat flies (Diptera: Streblidae and Nycteribiidae) on Neotropical bats (Chiroptera). Can J Zool 82:616–626

Theodor O (1957) Parasitic adaptation and host-parasite specificity in the pupiparous Diptera. In: Mayr E (ed) First symposium on host specificity among parasites of vertebrates. Institut de Zoologie, Université de Neuchâtel, Switzerland, pp 50–63

Timm RM (1987) Tent construction by bats of the genera *Artibeus* and *Uroderma*. In: Patterson BD, Timm RM (eds) Studies in neotropical mammalogy: essays in honor of Philip Hershkovitz, vol 39, Fieldiana: zoology, new series . Field Museum of Natural History, Chicago, pp 187–212

Timms R, Read AF (1999) What makes a specialist special? Trends Ecol Evol 14:333–334

Tompkins DM, Clayton DH (1999) Host resources govern the specificity of swiftlet lice: size matters. J Anim Ecol 68:489–500

Trowbridge RE, Dittmar K, Whiting MF (2006) Identification and phylogenetic analysis of *Arsenophonus*- and *Photorhabdus*-type bacteria from adult Hippoboscidae and Streblidae (Hippoboscoidea). J Invertebr Pathol 91:64–68

Wenzel RL (1976) The streblid batflies of Venezuela (Diptera: Streblidae). Brigham Young Univ Sci Bull 20:1–177

Wenzel RL, Tipton VJ (1966) Some relationships between mammal hosts and their ectoparasites. In: Wenzel RL, Tipton VJ (eds) Ectoparasites of Panama. Field Museum of Natural History, Chicago, pp 677–723

Wenzel RL, Tipton VJ, Kiewlicz A (1966) The streblid batflies of Panama (Diptera: Calypterae: Streblidae). In: Wenzel RL, Tipton VJ (eds) Ectoparasites of Panama. Field Museum of Natural History, Chicago, pp 405–675

Whitaker JO Jr, Ritzi CM, Dick CW (2009) Collecting and preserving ectoparasites for ecological study. In: Kunz TH, Parsons S (eds) Ecological and behavioral methods for the study of bats. Johns Hopkins University Press, Baltimore, pp 806–827

Whiteman NK, Sánchez P, Merkel J, Klompen H, Parker PG (2006) Cryptic host specificity of an avian skin mite (Epidermoptidae) vectored by louseflies (Hippoboscidae) associated with two endemic Galapagos bird species. J Parasitol 92:1218–1228

Whiting MF, Whiting AS, Hastriter MW, Dittmar K (2008) A molecular phylogeny of fleas (Insecta: Siphonaptera): origins and host associations. Cladistics 24:677–707

Chapter 7
No Myth But Reality: Blood Licking Bats

Heinz Mehlhorn

Abstract Bats (Chiroptera) are flying mammals, which occur worldwide. Since they are active at night, often occur in masses and appear as quickly moving shadows due to their very large wings, which are formed by skin that is stretched between their extremely enlarged fingers of the forelegs and the hind legs, they spread fear and horror since the times of the first human high cultures. They were perceived as flying malicious followers of the devil or other evil powers, although most bats never attack humans. Most bat species ingest insects during their flights. A few species eat fruits and thus may damage seriously fruit farms within one single night. Only three species attack animals (or rarely humans) when these potential preys are sleeping outside of houses, e.g., unprotected on meadows, etc. The males and females of these three species, which have their biotopes in the Americas (reaching from Chile/Argentina to Texas), have their touchdown close to sleeping animals, then crawl on the ground to their prey, scratch a small wound in the skin by help of their sharp and pointed teeth and finally lick the flowing blood. This is, however, the moment, when agents of diseases (e.g., viruses of rabies) could be transmitted. This chapter deals just with the blood licking species, while other bats are presented in separate chapters of the book.

Keywords *Desmodus rotundus* • *Diphylla ecaudata* • *Diaemus youngi* • Blood licking • Vampirism • Blood feeding mammals • Night activity • Vampire bats • Vectors of rabies

Three species of bats (Chiroptera, from *Greek* cheir = hand; pteron = wing) are known as so-called vampire bats to attack animals and occasionally humans:

H. Mehlhorn (✉)
Institute for Parasitology, Heinrich Heine University, Universitätsstr. 1, 40225 Düsseldorf, Germany
e-mail: mehlhorn@uni-duesseldorf.de

S. Klimpel and H. Mehlhorn (eds.), *Bats (Chiroptera) as Vectors of Diseases and Parasites*, Parasitology Research Monographs 5, DOI 10.1007/978-3-642-39333-4_7,

- *Desmodus rotundus* (common vampire)
- *Diphylla ecaudata* (whitewing vampire)
- *Diaemus youngi* (combteethed vampire)

As the other worldwide occurring bats they are mammals with a constant body temperature of 31 °C, give birth to their progeny and possess an excellent system of hearing reaching of up to 400 kHz, while the hearing of humans stop already at 21 kHz (Flindt 1986; Hoyt and Altenbach 1981).

These three blood engorging bat species have their habitats in warm regions that stretch from Chile/Argentina to Texas and are also found on some islands in the Caribbean Sea. They have a body length of 6–10 cm, but their total wing span may reach the respectable diameter of up to 40 cm, which is impressive when they fly at night with a speed of up to 50 km/h around humans (Flindt 1986; Siewing 1985). Their dorsal fur color is dark brownish, while the belly appears lighter (grey). The hind legs of vampire bats are rather strong compared to other bat species and thus they are able to walk or jump on the ground when approaching sleeping preys at night and when starting their flights from the ground. The vampire bats prefer special hosts. *Desmodus rotundus* feeds exclusively on mammals, *Diphylla ecaudata* exclusively on birds (Hoyt and Altenbach 1981), while *Diaemus youngi* sucks blood on both types of hosts. Having discovered a sleeping host, the bats land close by on the ground. A direct landing on a host would probably wake up this animal. Therefore, the vampire bats crawl or hop so close to their prey that they are able to bite. By help of temperature receptors around their mouthparts, they start to search for a blood vessel. At first, they lick at the selected skin places, whereby the anaesthetizing components of the saliva make them numb for the following bite. The bats bite by help of their pointed eye teeth and their sharp incisors. Thereby they scratch skin and thus reach the underlying blood vessels. Their saliva contains as mentioned above anticoagulants that keep the blood liquid (as it is in the case of leeches, too). By help of their grooved tongue, the bats lick the flowing blood until they have engorged around 30 ml. Having finished their meal within less than 2 min, they fly back to their sleeping places in caves of mountains, into mines, barns, or they even settle in rotting large trees. These vampire bats live together in rather small groups of 100 individuals (when e.g., compared to the hundreds or even thousands of tree-fruit bats in Australia). They have developed social system to help sick, weak, or unsuccessful members of their group. Successful hunters will share their ingested meal. After arriving at their sleeping places, they vomit—if necessary—some portions of their blood meal to feed weak members of the group.

Molecular biological studies gave hints that the development of blood feeding was only developed once within the group of bats (Tellgren-Roth et al. 2009; Wetterer et al. 2000). These findings are based on the genus specific activity of a few genes that help to express the anticoagulant components (i.e., so-called plasminogen activators (PA)—relatives of the group of serine proteases). It has been evaluated that feeding of bird blood is the most ancient way of food uptake, followed by enlarging their food spectrum from birds to mammals and finally reaching exclusivity in *Desmodus rotundus* of feeding on mammals after the genetic development of four different plasminogen activators (PAs) had been successfully developed.

The importance of vampire bites in humans and farm animals is rather high, since it was proven that the vampire's saliva may contain viruses that may introduce severe and possibly fatal diseases. Thus, death toll among humans due to the rabies virus disease is high in rural regions, the homeland of these bats. However, the transmission of other viruses is also possible and needs much more consideration. Since several hundred thousands of cattle are attacked per year, their growth and milk production is considerably reduced thus leading to large economic losses in these relatively poor countries in South and Central America. Therefore, there are yearly campaigns of eradication of the especially numerous bats of the species *Desmodus rotundus*, which is absolutely not endangered in contrast to many insectivorous bat species.

The mysteries around bats led to the creation of a human bat (i.e., Batman) first in comic books and later in movies (Banhold 2008; Beatty 2002), wherein the enormous flight abilities and the nocturnal activity of true bats are the predominant skills of this mysterious combination of human activity and animal behavior of Batman.

Simplified nomenclature of blood licking bats

Class	Vertebrata
Order	Chiroptera
Suborder	Microchiroptera
Family	Desmodontidae (true vampires)
Species	*Desmodus rotundus*
	Diphylla ecaudata
	Diaemus youngi

Origin of names:

- Chiroptera = bats. The name comes from *Greek* cheir = hand and pteron = wing
- *Desmodus* = *Greek* desmos = band; *Latin* modus = type; *rotundus* = Latin = round
- *Diphylla* = *Greek* dis = double; *Greek* phyllon = leaf; ecaudata *Latin* e = without, *caudatus* = with tail
- *Diaemus* = *Greek* di = two, *Latin aemulus* = imitating; *youngi* = name of a scientist (Young)

Rabies

This disease is based on infections with so-called ***Lyssa* viruses**, which belong to the Rhabdovirus group. Among the *Lyssa* virus group, seven different genotypes are described; five of them have their exclusive reservoir in worldwide occurring bats (Chiroptera), while genotype 1 (the "classic" agent of rabies) may also be transmitted by bites of several terrestic animals (e.g., foxes, dogs, and cats). The WHO claims that there are approximately 55,000 human death cases per year alone in Africa and Asia being induced by dog bites, while in Europe "only" 50 persons die caused by such rabies

(continued)

Rabies (continued)

infections. The bullet-like appearing rabies viruses measure 100–300 nm in length with a diameter of 75 nm and contain a helically arranged ribonucleocapsid with about 12,000 nucleotids.

The viruses are present in the saliva of infected animals and thus are transmitted by bites. Just after infection their reproduction starts in the cells of the surrounding tissues and/or in the local nerve endings. The **incubation period** of the disease is rather variable. In 25 % of the cases, the first symptoms start within the first 30 days, while in 50 % of the cases, it takes 30–90 days. In the rest of cases, the incubation period may be prolonged considerably.

The **clinical symptoms** are characterized by unspecific symptoms such as vomiting, dizziness, and headache. After this phase neurological symptoms follow. During this rather short period of 2–7 days, the following symptoms are seen: encephalitis, aggressive behavior, aero- and hydrophobia and the very typical hypersalivation. Death occurs finally due to breakdown of kidneys and of blood circulation.

If an infection and clear symptoms had occurred, there are no more chances for a causative medical cure. However, as long as symptoms have not yet occurred, there is the possibility of a so-called postexposition prophylaxis by help of a tissue cell vaccine (Roß et al. 2009). However, efficacy and healing is not guaranteed. Thus, it is very important that people working in forests or on farms become vaccinated, which is given in three doses regularly on days 0, 7, 21, or 28, so that beginning from day 35 protection is reached. Furthermore, it should be noted that wild animals that try to come into close contact to humans are mostly infected by rabies (Neumeister et al. 2009).

References

Banhold L (2008) Batman, 2nd edn. Bachmann, Essen

Beatty S (2002) Batman: Die Welt des dunklen Ritters. Dino, Stuttgart

Flindt R (1986) Biologie in Zahlen. G. Fischer, Stuttgart

Hoyt RA, Altenbach JS (1981) Observations on *Diphylla ecaudata* in captivity. J Mammal 62:215–222

Neumeister B, Geiss HK, Braun RW, Kimmig P (eds) (2009) Mikrobiologische Diagnostik. Thieme, Stuttgart

Roß RS, Müller T, Roggendorf M (2009) Rhadoviruses. In: Neumeister B, Geiss KH, Braun RW, Kimmig P (eds) Mikrobiologische Diagnostik, 2nd edn. Thieme, Stuttgart

Siewing R (ed) (1985) Lehrbuch der Zoologie, vol 2, Systematik. G. Fischer, Stuttgart

Tellgren-Roth A, Dittmar K, Massey SE, Kemi C et al (2009) Keeping the blood flowing – plasmine activator genes and feeding behavior in vampire bats. Naturwissenschaften 96:39–48

Wetterer JK, Rockman MV, Simmons NB (2000) Phylogeny of phyllostomid bats: data from morphological systems, sex chromosomes and restriction sites. Bull Am Mus Nat Hist 248:1–19

Chapter 8
Vampirism in Medicine and Culture

Peter Mario Kreuter and Heinz Schott

Abstract The aim of this chapter is multifold. Firstly, we want to show that the topic of vampirism is linked to a huge amount of misconceptions that have formed the idea of the vampire figure in Central and Western Europe for at least 100 years by analysing the role of the bat. Being a part of Southeast European vampire belief, this animal has no link to the vampire figure in Romania or Bulgaria. Secondly, the idea of vampirism in medicine and popular culture has to be explained. Finally, the link between vampirism, humoural pathology and magic medicine has to be enlightened.

Keywords Bat • Blood Transfusion • Folklore • Humoural Pathology • Magic Medicine • Medical History • Southeast European Folk Belief • Vampire

8.1 Some Remarks on the Vampire in Southeast European Folklore

The topic of vampirism is linked to a huge amount of misconceptions that have formed the idea of the vampire figure in Central and Western Europe for at least 100 years. Maybe the best example for this is the permanently repeated claim that Bram Stoker had a historical model in mind when writing his novel "Dracula". And so the Wallachian prince Vlad III the Impaler (1431–1477) became a permanent fixture in the Western popular literature about vampires in which he is always

P.M. Kreuter (✉)
Institut für Ost- und Südosteuropaforschung, Landshuter Straße 4, 93047 Regensburg, Germany
e-mail: kreuter@ios-regensburg.de

H. Schott
Medizinhistorisches Institut der Universität Bonn, Sigmund-Freud-Straße 25, 53127 Bonn, Germany
e-mail: heinz.schott@ukb.uni-bonn.de

S. Klimpel and H. Mehlhorn (eds.), *Bats (Chiroptera) as Vectors of Diseases and Parasites*, Parasitology Research Monographs 5, DOI 10.1007/978-3-642-39333-4_8,

presented as the historical inspiration for Stoker's vampire count. Nothing could be more wrong than that. Elizabeth Miller, a Canadian specialist for the œuvre of Bram Stoker, underlined that Stoker did not know very much about Vlad the Impaler (Miller 2005). In fact, it was just his other nickname "Drăculea", which attracted him by its sound and the strange visual effect. Apart from that, Stoker did not know that much about this ruler of Wallachia because his main source for the history of nowadays Romania mixed up father (Vlad II) and son (Vlad III) calling them both just "Dracula".[1] The rest of the so-called historical background in the novel is merely literary invention by Stoker. Another common misconception about vampires is their predicted aversion to sunlight. It is true that Stoker's Count Dracula preferred to move around during the hours of darkness, but there are several examples in the novel where he is walking through London during daylight time. And the vampire from the Balkans may be inactive on a certain day of the week (notably Saturday or the feast of the village's saint), but sunlight itself is very rarely a danger. Like many other things, this part of the westernised vampire was invented by the cinematic adaptations of Stoker's novel (1897).[2]

Among those misconceptions can be found that of the role of the bat in the popular Southeast European vampire belief. At a first glance, this may surprise not only by the fact that there are some bat species sucking (or better, licking and drinking) blood but also by the general existence of the bat in the folklore in Southeast Europe. But if we take a closer look at the popular vampire belief, one can understand that the bloodsucking bat could not under any circumstances be a part of the popular belief of that part of Europe.

So let us start with vampire figure in the Balkans itself. Or maybe it is better to say "with the vampire figures," as we can clearly note that there are so many different variations of that kind of revenant that it is hard to bring them down to a more general definition. Already the name for the vampire may vary not only from country to country—"vampir" in the countries of Slavic language, "lugat" in the Albanian speaking zone or "strigoi" in Romania—but also within a country or even a region. In Bulgaria, we have a lot of similar sounding words like "fampir", "vapir", "voper" or "vipir", changing from village to village and used to replace the taboo word "vampir", in Albanian we find also "dhampir", and in Romania "strigoi" is partially replaced by "moroi" or "bosorcoi".[3] Like the denomination, the characteristics of the vampire figure may be full of varieties. Just to mention one as an example—there is absolutely no structured geographical distribution regarding the question whether or not a vampire is able to speak. And even the kind of wood that should be used for the stake to be driven through a vampire's heart cannot be answered without the exact definition of the region one is actually talking about. If we reduce the vampire to his very basics, we can only state that he is a dead person returning as such from the grave because he was not able to go over into the

[1] Miller (2005), 112f.

[2] Miller (2005), 45.

[3] Kreuter (2001), 68f.

other world. And by not belonging either to this world where he comes from and is now trapped in, the vampire causes trouble, destroys the quality of corn or fruit, mixes up and bothers the cattle or brings even diseases or maybe death to the humans. But one thing he never does—biting the humans into the neck in order to suck their blood. This element of the vampire is pure invention of novels and movies and has no origin in the popular vampire belief in the Balkans. Well, in some cases, one can read or hear people making statements like “and then the vampire came and took his blood”. But when asking how he did it, no answer can be given because he takes the blood, and that’s it. Any more detailed idea does not exist, and especially the possibility of biting and sucking is never ever mentioned or at least thinkable. It is just a description, a metaphor. And quite rare.

8.2 The Vampire and the Bat

Returning to the bat and its role in the Balkanic folklore, one has to state that this animal is not among those with the highest importance in the popular belief. Of course, there are traditions and believes in connection with the bat, but they are far away from any relation to blood and not linked to the vampire figure. And the bat’s position is very ambiguous, on the one hand a demonic animal, on the other hand linked to joy and good luck. “As among the Germans of Transylvania and the Upper Palatinate, bats were interpreted by the Albanians as omens of death. Thus, if a bat flew into a house, it was thought that someone there would die”,[4] is a typical interpretation of the demonical function of the bat. A similar belief among the South Slavs is linked to diseases, especially to the plague, which is thought to come to a village in the form of a bat,[5] even though for the Bulgarians we can state that also certain kinds of butterflies can be the bringers of the plague.[6] Very common is a connex made between the bat and hair. The hair of a bat could be used in parts of Bosnia and Herzegovina for love potions if three small bat hairs are poured into the coffee of the loved one.[7] As the bat is seen as a blind, this animal can also be used to make a boy blind of love, if the girl is walking around him three times with a bat disguised in her clothes.[8] But then the bat brings also good luck and joy. In and around Sarajevo for example, bats living under the roof of a house will bring good luck to the family.[9] These examples show clearly that there are elements in Southeast European folklore dealing with the bat in some ways. But one thing is missing—a link to the vampire. The vampire of Serbia, Romania or Bulgaria is also

[4] Elsie (2001), s.v. ‘Bats’, 24.

[5] Bächtold-Stäubli (1930), s.v. ‘Fledermaus, col. 1590.

[6] Dukova (1997), 30f.

[7] Bartels (1899), 249.

[8] Ibidem, 249.

[9] Ibidem, 254.

a shape-shifter, he is able to change his exterior into that of a wolf, a horse, a dog, a bird or a butterfly, but the bat is not mentioned in the reports and interviews made by folklorists. And as there are neither bloodsucking revenants nor blood-drinking bats in the Balkans, the idea of the "vampire bat" was unable to rise.

But how came the blood-drinking bats of South America to their vampiristic names and images? This is due to the widespread success the word "vampire" itself had after 1732 in the European languages. When Austrian military surgeons investigated from 1725 on several cases of pretended vampire attacks in Bosnia and Serbia, they wrote also official reports about their investigations. These reports in which the idea of blood sucking is falsely mentioned became known among the scientists of Europe through articles in newspapers, and by the reception of those journals and pamphlets within a large Enlightened debate between 1732 and 1739, the word "vampir" made its way into all languages of Central and Western Europe.[10] Years later, the French naturalist Georges-Louis Leclerc, Comte de Buffon (1707–1788) labelled a bat scientifically known as Desmodontidae—vampire bat. From 1749 on, he published the "Histoire naturelle, générale et particulière" (36 volumes till 1788; an additional volume based on his notes appeared in 1789), in which he used the testimonies of travellers and naturalists, as well as his own experience. And as he implied that these bats suck the blood of humans, he named them after the vampires of Southeast Europe.[11]

8.3 Vampirism and Medical History

But what have vampirism and medical history in common and in which way they have different views? First of all, it is remarkable that in vampirism the boundary between life and death is partially abolished. This is a provocation for medicine being forced to define diagnostically whether a body is really dead or still alive. The traditional task of medical doctors was to exclude apparent death. This was a real problem until the early nineteenth century, when strict rules for necropsy were introduced. The fear of being buried alive produced bizarre devices of alarm on graves.[12] The idea of a "living corps", which was actual throughout the whole history of culture, was not tolerable for modern scientific medicine. So, the brain death diagnostics introduced about 1970 was necessary for the transplantation of (vital) organs the owner of which had to be declared dead before the explantation. There is an interesting analogy between Vampirism and medicine in regard to epidemic infection. The vampire can infect healthy persons and produce epidemics especially in form of bloodsucking bugs. Probably certain epidemics in the Balkan countries in the early eighteenth century had their impact on the origin of the

[10] Kreuter (2012), 13–22.

[11] Kreuter (2001), 31f.

[12] Schott (2005), cols. 678–683.

vampire delusion—a problem for the contemporary learned doctors to disenchant the vampire myth in the sense of the ongoing enlightenment (Grothe 2001).

8.4 Bleeding and Blood Transfusion: Vampirism and the Humoural Pathology

In the history of medicine, blood is the most important vital and therefore its red colour became the attribute for medical faculties, e.g. in regard to robes or the cover of scientific series. Blood was—besides yellow bile, black bile and phlegm—one of the four cardinal fluids or humours according to the antique humoural pathology. It was thought to have warm and dry quality; it represented dawn (aurora), childhood, spring and moreover sanguine temper originating from the heart. In general, blood symbolised the life force animating physiologically all the organs. Blood loss in a healthy body implied a vital danger, whereas bloodletting in a sick body would cure it. According to the humoural pathology corrupted fluids (humours), especially blood poisoning, would cause all possible diseases, which had to be cured by therapeutic methods of discharge.

Apart from this, there are two further medical aspects of the vampirism. On the one hand, there is a polarity between health and disease and on the other hand one between the doctor and the patient. The sucking of blood by a vampire has regarding the aspect of health and disease a double effect: It supplies the vampire with vital energy healing him by a perverse action, in contrast to the victim whose vital energy is extracted causing weakness and disease. So, a complementary relation of weakening and strengthening is established. Regarding the relation of the doctor and his patient two different procedures may occur: a good doctor helps his patient to resist all attacks of weakening powers preserving him from the loss of vital energy, whereas a bad doctor himself is a sort of vampire exhausting his patient, in particular financially.

Bloodletting was a classical therapeutic method practised since antiquity. But insofar, it was discarded the analogy to the vampirism is not quite correct; it resembled more the kosher butchering of animals. As far as I know, there was no case of a doctor really drinking the patient's blood or giving the blood to another one. But in the antiquity, fresh human blood was appreciated as a medicine and an agent of rejuvenation. So, the Roman author Plinius the Elder (1st c.) recommended to drink the blood of wounded persons, especially that of gladiators, as an aid against the falling sickness.[13] The drinking of the blood of children is mentioned in the literature again and again. According to the legend Pope Innocent VIII (1432–1492) is supposed to have drunken the blood of three young children to rejuvenate himself. It is also said that the famous Renaissance philosopher Marsilio Ficino, a contemporary of the Pope, recommended drinking the blood of young

[13] Schott (1993), 6.

people mixed with sugar as an agent of rejuvenation. Analogously, Ficino recommended finding a healthy pretty girl to suck on breast during full moon.[14]

The early modern blood transfusion displays an interesting aspect. The first transfusions of animal blood were performed after the invention of the intravenous injection technique in the 1660s in Western Europe. In 1667, the French mathematician Jean Baptiste Denis and his compatriot, the surgeon Paul Emmerez transfused lamb blood to a 15-year-old patient suffering from fever who was weakened by repeated bloodlettings. The apparent success was followed by unavoidable horrible out comings, the "sheep melancholy" as a consequence of the transfusion of lamb blood. It was questioned, whether with the blood also characteristic traits would be transferred. So, one speculated whether a sheep would become biting like a dog or vice versa whether a dog would develop wool and horns infusing the blood of a sheep. The German physician and alchemist Johannes Sigismund Elsholtz recommended even in his textbook "Neue Clystierkunst" the blood transfusion between humans to reconcile quarrelling brothers or spouses.[15] In this perspective, the drinking of blood for the accomplishment of a blood brotherhood can be understood as a sort of mutual vampirism to produce a sympathetically common sense.

The application of leeches was a common method of discharging in humoural pathology. The leech as a bloodsucker symbolised vampirism very impressively. Whereas the critics of the bloodletting blamed this operation as a dangerous vampirism, its supporters highlighted it as their therapeutic guideline. Because of their excessive discharging procedures many barber surgeons appeared like vampires in the early modern period taking off not only the blood but also the money from their patients cupping them in this double respect.

8.5 The Extraction of Life Force: Vampirism and Magic Medicine

In medicine, blood was traditionally linked with the idea of the life energy or life force of an organism. The concept of the life force (vis vitalis, German: Lebenskraft) was differently designed, e.g. as a nerve spirit (spiritus nervosus), a life spirit (spiritus animalis), archeus, fluidum, etc. The respective concepts founded the early modern magic medicine assuming magnetic–sympathetic interactions between corresponding natural things including living and dead humans. From the antiquity there is the idea, that life energy can be extracted from the human organism by certain practices producing diseases and even death. A complementary idea is the opposite procedure: to transfer of life energy to a weak organism to empower and to cure it. A special meaning has the combination of both

[14] Illustrierte Geschichte der Medizin (1982), 2096.

[15] Ibidem, 188.

procedures: the extraction of life energy out of the human body and its transfer to another one.

Paracelsus described this double procedure in his tract "Herbarius" (written before 1527). In the chapter on the "English Thistle" (Angelica, Englische Distel), a well-known traditional medicinal plant, he writes: This plant of the English thistle has its natural effect similar to the moon, it extracts the power from somebody and transfers it someone wearing it at the same time".[16] Paracelsus stressed that this was a natural process. If somebody does not use his strength, it should be taken from him and transferred to another one. He reported an example: "Once I saw a man in Alsace bearing three zentner wine in a barrel from Ruffach to Sulz a long mile away fixed on him, who was accompanied by 12 men and tired them walking, so that they could not follow him and became weak and some of them lay down weakened.[17] As Paracelsus explained, that the trial failed curing a sick person by surrounding him with strong people and giving him *Angelica*. This would help only when somebody worked very hard. The character of the sponsor would transfer together with his force: "it is also within himself, when a force is extracted from another one, so that the person who gets it, follows it. Is the other one a choleric, the host will it also be..."

The "gerocomics" (from Greek *geron* = old and *komeo* = to care) is a prominent example for the magical transfer of life energy throughout the ages. It was practiced since antiquity and still discussed intensively in medicine during the romantic era in the early nineteenth century. So, the German doctor Georg Friedrich Most wrote in is textbook "The sympathetic means and healing methods" (Die sympathetischen Mittel und Curmethoden): "One knew even in the antiquity, that living animal bodies have the ability to transfer a part of their plenty of life to others, which was fundamental for the art of gerocomics rejuvenating worn out old men by the close atmosphere of a fresh young person. That the latter are weakened by this being together is witnessed by Galen, Baco, Sydenham, van Swieten, Boerhaave and also Wurzer, Gmelin and Kluge based on own observations".[18] The respective conceptions of the "effectiveness of the life vapour" (Wirksamkeit des Lebensdunstes) and the "capacity of a sensible body atmosphere" (Wirkungsvermögen einer sensiblen Körpersphäre) were discussed about

[16] "Dise wurz des englischen distels ist dermaßen in ir natur, so sie in irer operation ist, welche gleich dem mon ist, so zeucht sie von einem andern die kraft aus und gibts dem, der sie tregt zu der selbigen zeit." Cf. Theophrast von Hohenheim (1930), 35: The Angelica was a Nordic drug listed in the pharmacopoes from tz sixteenth untul the twentieth centuries, originally recommended as a plague, later as a gastric remedy.

[17] "!ch hab erst mal gesehen, das ein man im Elsaß tragen hat von Ruffach gen Sulz drei zentner schwer ein lange meil wegs wein in einem faß auf sich gebunden, und 12 man zu im genomen, hat die 12 alle müd gangen, das sie im nicht haben mögen folgen und schwach hernach gangen, etlich tag hernach gar geschwecht gelegen". Ibidem, 38 seq.

[18] "Schon im Alterthum wußte man, daß der lebende Thierkörper die Eigenschaft besitze, einen Theil seiner Lebensfülle auf Andere zu übertragen, worauf sich ja die Kunst der Gerocomik stützte, um abgelebte Greise, durch die nahe Atmosphäre einer frischen Jugend wieder zu verjüngen. Daß letztere durch dieses Beisammenseyn aber geschwächt werden, dies bezeugen schon Galen, Baco, Sydenham, van Swieten, Boerhave [sic] auch Wurzer, Gmelin und Kluge, gestützt auf eigene Beobachtungen". Most (1842), 114 seq.

1800 particularly within the context of animal magnetism or mesmerism as it was later called. The above-mentioned Gmelin and Kluge were well-known promoters of mesmerism in Germany.

Even Christoph Wilhelm Hufeland (1797), who was a critic of animal magnetism, mentioned the gerocomics as a "practice to rejuvenate and conserve old worn out bodies by the close atmosphere of a fresh flourishing youth" (die Gewohnheit, einen alten abgelebten Körper durch die nahe Atmosphäre frischer aufblühender Jugend zu verjüngen und zu erhalten). Even recently, this practice would have been used "with benefit": "the great Boerhaave let sleep an old lord mayor of Amsterdam in between two young people [two young and virtuous women[19]] and states that the vivacity and powers of the old obviously had increased. And certainly, considering the influence of the life vapour of animals freshly cut open on palsied limbs these methods does not seem condemnable".[20] About 1800 the experimentations with animal magnetism boomed and many experiences and phenomena of an apparent transfer of "fluidum", as the subtle agent was called, were reported. An illustrious example should be mentioned here. The Swabian doctor and poet Justinus Kerner hosted a special patient in his household for about two and a half years: Friederike Hauffe (1801–1829), "the Seeress of Prevorst" (die Seherin von Prevorst) suffering from very severe somatic and psychic disorders getting strange experiences during her somnambulant states of consciousness. Kerner (and sometimes other members of his family) applied often the magnetic cure. His son Theobald described his own experience with the "Seeress": "Often, when my father had to visit a patient far away and could not magnetise the Seeress at the usual time, he magnetised me before his departure, and when I charged with this subtle fluidum met her at the anticipated time I was especially welcome. I had to sit down on her bed calmly, she took my hand, and I had to stay motionlessly until she had absorbed the fluidum commissioned to me, her eyes closed, her hands relaxed; then I got up silently, slipped through the door, and avoided to see again this spider sucking my nervous power".[21] These sessions and Samaritan services for the Seeress had often an evil effect. But also Justinus Kerner himself observed a similar phenomenon interacting

[19] Illustrierte Geschichte der Medizin (1982), 2096.

[20] "Der grosse Boerhaave liess einen alten Amsterdamer Bürgermeister zwischen zwey jungen Leuten [zwei jungen und tugendhaften Frauen] schlafen, und versichert, der Alte habe dadurch sichtbar an Munterkeit und Kräften zugenommen. Und gewiss, wenn man bedenkt, was der Lebensdunst frisch aufgeschnittener Thiere auf gelähmte Glieder, was das Auflegen lebendiger Thiere auf schmerzhafte Uebel vermag, so scheint diese Methode nicht verwerflich zu sein". Hufeland (1797), 10 seq.

[21] "Gar häufig, wenn mein Vater über Feld zu Kranken mußte und die Seherin nicht zur gewohnten Stunde magnetisieren konnte, magnetisierte er mich vor seiner Abreise, und trat ich dann, mit diesem unwägbaren Fluidum beladen, zu angegebner Zeit bei ihr ein, so war ich besonders willkommen. Ich mußte mich ruhig an ihr Bett setzen, sie ergriff fest meine Hand, und ich mußte unbewegt ausharren, bis sie das mir anvertraute Fluidum aufgesogen hatte, ihre Augen sich schlossen, ihre Hände sich lockerten; dann stand ich leise auf, schlüpfte zur Tür hinaus und ließ mich womöglich den ganzen Tag nicht mehr bei der an meiner Nervenkraft saugenden Spinne sehen". Kerner (1983), 34.

with the Seeress: "Mainly she absorbed Pabulum vitae [that means a nervous spirit; "Nervengeist"] from the eyes and finger tips of stronger persons which often did not feel it, but often did it very much".[22] This reminds us of the concept of "Od vampirism" (Od-Vampirismus),[23] which was coined in the 1840s in regard to the at that time spectacular of doctrine of the Baron Karl von Reichenbach, a successful chemist, entrepreneur and also natural philosopher.

We may formulate a general hypothesis: The suggestive power of the vampire belief may produce a severe shock syndrome and even sudden death, because there is probably no image more dangerous than to bleed to death. At the end of the nineteenth century, US-American doctors made a weird experiment to prove the power of suggestion. The test subject was somebody sentenced to death delivered by the legal authority. They told him that they would execute him by opening his jugular vein. They blindfolded him, scratched slightly the skin, and let flow down lukewarm water, which was caught in a bowl. Within a few minutes the man was dead.[24]

8.6 Vampirism as a Metaphor: On the Social Stigmatisation of "Evil" Persons

The idea of vampires played also a role in the imagery of medicine in regard to social and political aspects. Vampires symbolised in a way parasites within the "popular body" (German: "Volkskörper") exhausting and damaging it as social spongers. In the eighteenth century, the blood was focussed by the humoural pathology and became the main fluid (humour). Remarkably, "blood" became the key concept of the race biology in the late nineteenth and early twentieth centuries. In the Nazi ideology ("Blut und Boden"), the pure blood of the Aryan race had to be protected from foreign contaminations, especially by "Jewish blood". The Nazi "Law for the Protection of German Blood and German Honour" (Gesetz zum Schutze des deutschen Blutes und der deutschen Ehre) in 1935 penalised the sexual contact with Jews. In Sect. 8.2 it was declared: "Extramarital sex between Jews and nationals of German or allied blood is illegal". Such blood imagery implied further metaphors correlating with vampirism: The so-called antisocial people and especially the Jews as parasitic bloodsuckers, bugs, exhausting the healthy body of the people. The "Jewish blood" seemed to threaten the German people by a final blood

[22] "Hauptsächlich sog sie [den 'Nervengeist'] aus Augen und Fingerspitzen anderer stärkerer Menschen, von diesen oft nicht gefühlt, auch oft sehr gefühlt, ein Pabulum vitae [Lebensfutter] in sich". Kerner (1829), 175.

[23] Spiesberger (1953), 64.

[24] Bechterew (1905), 35.

poisoning. The greatest danger would originate from it and alienate the Aryan race.[25] The anti-Semitism ascribed two traits to the Jews characteristic for the vampirism too: firstly the absorption of blood respectively vital power, and secondly the infection of a healthy population (folk) with a poison or germ damaging the vital power.

8.7 Vampirism and Esotericism: Some Remarks

Within the scope of the contemporary esotericism, especially in the field of Satanism, the vampirism appears apparently as a model for acquiring new life energies. There is a "website für paganismus" explaining under the headline: "Magic Vampirism": "I considered often practising a sort of magic vampirism, to upload my creative power with new energy. When I feel 'charged' I can perform my art considerably more easily and can produce faster and have generally more perseverance in my work. But I have such a demand of psychic power that I cannot cover the needed demand. [. . .] What are the experiences you have made with 'magic vampires', and is there a constructive possibility to combine the power of several gifted users? Are there methods of fast detection of potential and compatible 'hosts'?"[26] The same author wrote in another thread on the topic of "mental energy deficit": "I feel myself constantly weary, powerless, and grey. Is there such a thing like an auxiliary generator for this sort of power. [. . .] My projects on this field proceed only slowly. [. . .] Or is there a possibility to tap an external natural source? Perhaps similar to an incubus, I don't mean the literal translation but the respective essence in the infantile myths?"[27]

---

[25] Schott and Tölle (2006), 189.

[26] "Ich habe mich schon sehr oft mit dem Gedanken auseinandergesetzt, eine Art magischen Vampirismus zu betreiben, um meiner Schaffenskraft neue Energie zukommen zu lassen. Wenn ich mich 'aufgeladen' fühle, fällt mir meine Kunst erheblich leichter und ich kann schneller produzieren und bin im Allgemeinen ausdauernder bei meinen Vorhaben. Ich habe bloss einen so gewaltigen Bedarf an Seelenkraft, dass ich das benötigte Volumen nicht mehr selbst decken kann. [. . .] Welche Erfahrungen habt ihr mit 'Magie-Vampiren', und gibt es vielleicht auch eine konstruktive Möglichkeit die Kraft mehrerer begabter Nutzer zusammenzuschliessen? Und gibt es ein Schnellerkennungsverfahren für potentielle und kompatible 'Wirte'?"; cf. http://www.paganforum.de/magie-allg/4672-magischer-vampirismus.html ("Die Seite für das Heidentum"; i.e. "the website for the paganism"; 16.03.2013).

[27] "Ich fühle mich dauernd abgeschlagen, kraftlos und grau. Gibt es sowas wie einen Hilfsgenerator für diese Art von Kraft. . . meine Projekte auf diesem Gebiet kommen, aufgrund des eben genannten Problems nur langsam voran. . . Oder gibt es eine Möglichkeit eine Externe [sic] natürliche Quelle anzuzapfen? Ich meine vielleicht so ähnlich wie ein Incubus, ich meine nicht die wörtliche Übersetzung, sondern dieses Wesen in den infantilen Myten [sic]?" Ibidem, 16.03.2013.

8.8 A Final Remark: Vampirism and Everything

We started our overview by discussing the place of the bat in the popular belief of Southeastern Europe, and we ended up with the use of "pure blood" for the stigmatisation of "others" and vampirism in the esoteric movement. How can that be?

One thing is very important to underline—the vampire figure (and vampirism as a concept) is a perfect projection screen for a large number of different topics. The vampire does not need a mad scientist to get created, he is just there, existing by factors that are natural ones or caused by other human beings without any intention. This figure is put together by a number of different elements fitting well to a lot of different moments in a man's life or a village's situation. He could be made guilty for a large variety of things. And vampirism as a concept is understandable in nearly all cultures over the world. Bring this together with the blood and its importance not only for the vital function of the human body but also in the conceptions of a huge number of religions and the perfect "shape-shifter" of a mythological figure is born.

Thus, the vampire can be used for nearly everything. Even to give his name to a bloodsucking bat thousands of miles away from the Balkans.

References

Bächtold-Stäubli H (ed) (1930) Handwörterbuch des deutschen Aberglaubens, vol 2. Walter de Gruyter & Co., Berlin

Bartels M (1899) Ein paar merkwürdige Kreaturen. In: Zeitschrift des Vereins für Volkskunde, vol. 9, pp 171–179 und 245–255

Dukova U (1997) Die Bezeichnungen der Dämonen im Bulgarischen. Otto Sagner Verlag, München

Elsie R (2001) A dictionary of Albanian religion, mythology, and folk culture. New York University Press, New York, NY

Grothe S (2001) Der Einfluß der Seuchen auf die Entstehung des Vampirglaubens im Spiegel der Leipziger Vampirdebatte 1725–1734. Med. Diss., Köln

Hufeland CW (1797) Die Kunst, das menschliche Leben zu verlängern. Akademische Buchhandlung, Jena

Illustrierte Geschichte der Medizin (1982) vol. 6. Andreas & Andreas, Salzburg

Kerner J (1829) Die Seherin von Prevorst. Eröffnungen über das innere Leben des Menschen und über das Hereinragen der Geisterwelt in die unsere, 1. Theil. Cotta, Stuttgart

Kerner T (1983) Das Kernerhaus und seine Gäste. Eine Auswahl. Salzer, Heilbronn

Kreuter PM (2001) Der Vampirglaube in Südosteuropa. Studien zur Genese, Bedeutung und Funktion. Rumänien und der Balkanraum. Weidler, Berlin

Kreuter PM (2012) Vom volkstümlichen Vampir. Nebst einer kurzen Betrachtung zu Kindern in den Vampirberichten des 18. Jahrhunderts. In: J Mikota, S Planka (eds) Der Vampir in den Kinder- und Jugendmedien. Weidler, Berlin, pp 13–22

Miller E (2005) A Dracula handbook. Xlibris Corporation, Philadelphia, PA

Most GF (1842) Die sympathetischen Mittel und Curmethoden. Stiller, Rostock

Paganforum. http://www.paganforum.de/magie-allg/4672-magischer-vampirismus.html. Accessed 16 Mar 2013

Schott H (1993) Die Chronik der Medizin. Chronik-Verlag, Dortmund
Schott H (2005) Scheintod. In: von Jagow B, Steger F (eds) Literatur und Medizin. Ein Lexikon. Vandenhoeck & Ruprecht, Göttingen, cols. 678–683
Schott H, Tölle R (2006) Geschichte der Psychiatrie. Krankheitslehren – Irrwege – Behandlungsformen. Beck, München
Spiesberger K (1953) Justinus Kerners'Seherin von Prevorst' in der Betrachtung esoterischer Tradition und im Lichte moderner psychischer Forschung. In: Sopp E, Spiesberger K (eds) Auf den Spuren der Seherin. Osiris Verlag, Sersheim
Stoker B (1897) Dracula. Archibald Constable and Company, London
Theophrast von Hohenheim gen. Paracelsus (1930) Sämtliche Werke, hg. von K Sudhoff, 1. Abt., vol. 2. Oldenbourg, München
von Bechterew W (1905) Die Bedeutung der Suggestion im sozialen Leben. Bergmann, Wiesbaden

Chapter 9
Unsolved and Solved Myths: Chupacabras and "Goat-Milking" Birds

Heinz Mehlhorn

Abstract Since the years 1992–1995 until now, a still growing new myth (urban legend) arose from findings of dead and apparently bloodless animals close to human dwellings in the regions spreading North from Bolivia to the South of Texas with especially many cases in Mexico, Puerto Rico and on the Caribbean Islands. Many of the killed animals were goats and sheep. Thus, these mystic killers were called chupacabras, the name of which comes from the *Spanish* words chupar = to suck and cabra = goat. Since the murdering occurred at night, many myths arose to explain these killings. When finding remnants of strange looking degrading bodies, they were kept for chupacabras and in many cases explained as remnants of "aliens" visiting us from other worlds or as mystic animals mostly reaching a length of up to about 1.5 m. Rational explanations, however, were not accepted by the local people, where "voodoo" is still common practice. This wish to believe in a miraculous witch world is not restricted to the population of the New World but has also been common in times of the Greek and Romans as well as in the centuries of the so-called "dark Medieval" in Europe. Therefore, not only nightly active animals like bats but also silently flying birds like owls or nightjars were believed to be helpers of the devil. Thus, a harmless insect-feeding bird like the nightjar (*Caprimulgus europaeus*) got the trivial genus name "goat-milking bird," which comes from *Latin* capra = goat and mulgere = to milk and was thus feared for many centuries (although no more nowadays).

Keywords Modern urban legends • Chupacabras • Goat killers • Strange animals • Aliens • Mutilations • Modern legend • Case findings from Bolivia until Texas • Legend of milk-sucking birds • Nightjars • *Caprimulgus* species

---

H. Mehlhorn (✉)
Institute for Parasitology, Heinrich Heine University, Universitätsstr. 1, 40225 Düsseldorf, Germany
e-mail: mehlhorn@uni-duesseldorf.de

S. Klimpel and H. Mehlhorn (eds.), *Bats (Chiroptera) as Vectors of Diseases and Parasites*, Parasitology Research Monographs 5, DOI 10.1007/978-3-642-39333-4_9,

9.1 Chupacabras

Humans languish for news and especially for seemingly unexplainable phenomena and thus the modern world of media and especially the yellow press provides daily new stuff. Thus, the phenomenon "chupacabras" and its different explanations spread rapidly around the world, after in the years 1992–1995 reports from dead, apparently bloodless goats (being killed in the night) had their origin in a village called *Moca* being situated on the island Puerto Rico in the Caribbean Sea, where "voodoo"-practice is still common (Corrales 1997; Schneider 2008; Horn 2002; McNab 2007). At first, there had been found seemingly bloodless goats and thus the name "chupacabra" (from *Spanish* chupar = suck and cabra = goat) was created after the first reports in newspapers had been titled "El vampira de Moca." The news of this phenomenon was spread all over the world by articles in daily newspapers and by broadcasts of radio and television stations. Thus, the number of findings of such strange animals increased constantly in the countries from Bolivia to Texas. One side increasing numbers of dead house animals (dogs, goats, sheep, cattle, etc.) were found but also other animals, respectively organisms, which were in such an enlarged status of degradation that their systematic position in the animal kingdom was not easy to find out just by inspection with the naked eye.

The first group contained apparently the bloodless prey animals and the second obviously contained the unknown predators. When studying the individuals of the second group by means of molecular biology, it turned out that these had been coyotes or dogs, which had lost their hair due to an infection with scabies mites and/or due to other diseases. In addition, they were mostly completely dried out due to the heat in these regions. The animals, which in all probabilities belonged to the prey group, often supported also the impression to be bloodless. This was, however, mostly explainable by the fact that the dead bodies had also been exposed to the heat in those warm countries for a long time leading to coagulation of the blood cells in the blood vessels. Furthermore, it had been proven that in many cases animals of the prey group had been obviously manipulated by humans, who cut off pieces from blood vessels—a phenomenon which is known as "mutilation," whereby some crazy humans try to joke with others being driven by envy to damage the goods of others or by the wish to initiate public sensations.

Rational explanations (Schneider 2008; Horn 2002; McNab 2007), however, were mostly not accepted and the urban legend "chupacabras" is still today expanding, since the group of fans of aliens is considerably increasing, especially in the USA. These activities explain that the "chupacabra" creatures are either dead aliens or unknown and strange creatures that had been brought by aliens onto earth and which had been left behind, when these aliens started their return flights. Other legends have their background in the believe that the chupacabras have their origin in failed gene experiments done in secret laboratories, which belong either to governmental Secret Services or to world-threatening criminal organisations famous from James Bond movies. Thus, rational approaches to hardly explainable phenomena are much more difficult for many people, while the belief in conspirative plots or in transcendent events is of course much easier—especially in societies, where many people believe in miracles or neglect even today the evolution on earth.

9.2 Goat-Milking Birds: Nightjars

Not only during the centuries of the science-neglecting "dark Medieval" in Europe but also in the societies of the more enlightened Greek and Roman scientists, some nightly active animals were considered as dangerous and harmful to humans and/or their farm animals. Of course, at the first place of evil bringing animals ranged the bats, which often occurred hunting in swarms and showed pointed teeth, when found dead on the ground or hanging at animals, since some species are indeed blood lickers (see other chapters in this book). But also other harmless birds like owls or nightjars were considered with deep mistrust and gave cause for horror stories. Therefore, it is not astonishing, that the bird which is called nightjar today, got from Linnaeus (Linné) (1758) in his famous "Systema naturae" the scientific genus name *Caprimulgus* (from *Latin* capra = goat and mulgere = to milk). This name is based on a text of the Roman scientist Plinius (the elder one), wherein the belief was expressed that these birds suck at night the milk of goats, which thereby grow blind and/or die. The English name "nightjar" confers to the fact that the members of this and 6 other species are active during night and utter "jarring" voices during their otherwise silent flights. But also Linnaeus (1758) already knew that these night swallows are harmless insect feeders, which collect during the night insects for their brood, which must become strong enough within a short time in order to survive a long flight to their overwintering places in the South of the Sahara desert. These birds, which have their unequipped nests on the ground, got apparently their names and the blame that they milk goats and make them blind. This belief arose from the facts that these birds were often seen during the night close to herds of goats and that they are especially aggressive during ardour defending their territory. However, these birds were not attracted by the milk in the udder of the goats, but only by the occurrence of huge numbers of insects close to herds of goats, sheep, or cattle. On the other side, the shepherds had correctly observed that many animals on the meadows (especially in warm countries) suffered from blindness, reduction of milk production, or that they even died as young animals. Today, we know that the accused nightjars are innocent, since it is standard knowledge among parasitologists, veterinarians, and microbiologists that flies (such as *Oestrus ovis* and related species) place their eggs respectively larvae onto the eyes, which may destroy the eyes mechanically (Mehlhorn 2012a, b). Furthermore, the adult flies often transmit bacteria and viruses onto the eyes and eggs of parasites on nose and/or mouthparts, so that these animals get infected. Since the bacteria and/or the growing fly larvae lead to bad sight or even to blindness, it is clear that the goats or other herd animals feed less and thus reduce or even stop their milk production.

Additional worm infections also lead to reduction of body weight (Mehlhorn 2008, 2012a, b; Gestmann et al. 2012; Förster et al. 2012). Therefore, the observations of the presence of nightjars in the surroundings of herds on the meadow and the occurrence of blindness, reduction of milk production and/or loss of weight or even death led to the wrong conclusions in many former societies, which is finally blamed harmless wandering and insect feeding (=thus useful) birds as monsters. But even today, there are still many human populations, where superstition is existing—not only in the case of the "evil eye" in native populations but also in industrial societies, where the belief that a substance ingested after it had been diluted more than hundred thousandfolds is still unbroken and makes huge sums of money to go around. On the other side, there are tiny finches, e.g. on the Galapagos Islands, which have developed a strange behaviour: they pick at the basis of feathers of sleeping larger birds and lick droplets of blood.

Simplified system of nightjars	
Classis	Aves
Order	Caprimulgiformes
Family	Caprimulgidae
Genus	*Caprimulgus*
Species	*Caprimulgus europaeus*
	Caprimulgus ruficollis
	Other species and subspecies

Gajus Plinius Secundus (the elder) (23–79 after Jesus Christ)

He was an important writer besides his task as chief of the imperial fleet at Misenium. He died as victim of the outbreak of the volcano Vesuvius (close to Naples) in the year 79 after Jesus Christ. He is famous for his authorship of the "**Historia naturalis**," which consists of 37 volumes containing the knowledge of his time. Therein is also included the story that nightjar birds suck milk from goats.

References

Corrales S (1997) Chupacabras and other mysteries. Greenleaf, New York, NY

Förster M, Gestmann F, Mehlhorn H, Sievert K, Messler S, Neuhausen N, Petersdorf S, Pfeffer K (2012) Flies as vectors of microorganisms. In: Mehlhorn H (ed) Arthropods as vectors of emerging diseases, vol 3, Parasitology research monographs. Springer, Berlin, pp 227–254

Gestmann F, Förster M, Mehlhorn H, Sievert K, Messler S, Neuhausen N, Petersdorf S, Pfeffer K (2012) Flies as vectors of parasites. In: Mehlhorn H (ed) Arthropods as vectors of emerging diseases, vol 3, Parasitology research monographs. Springer, Berlin, pp 195–226

Horn RM (2002) In German: curious and phantastic form of life – vampires, mothmen, seasnakes, ghostdogs, Yetis, dragons and chupacabras. Bohmeier, Lübeck

Linné C (syn. Linnaeus) (1758) Systema naturae. L. Salvius, Stockholm

McNab (2007) In German: mystic monsters. The unbelievable creatures in legends, myths and media. Tosa, Fränkisch-Krombach
Mehlhorn H (2008) Encyclopedia of parasitology, 3rd edn. Springer, New York, NY
Mehlhorn H (2012a) Parasiten des Menschen, 7th edn. Springer Spektrum, Heidelberg
Mehlhorn H (2012b) Parasiten der Tiere, 7th edn. Springer Spektrum, Heidelberg
Schneider M (2008) In German: traces of the unknown. Kryptozoology, monster, myths and legends, 2nd edn. Twilight-Line, Krombach

Chapter 10
Myth and Reality: Candiru, the Bloodsucking Fish That May Enter Humans

Heinz Mehlhorn

Abstract Since the beginning of the eighteenth century, when European natural scientists travelled in the regions of the river Amazonas and its tributaries to study the new amazing world of plants and animals, several reports are available on the existence of tiny fish, which allegedly enter the sexual organs and the anus of female and male humans. However, an exact check of the literature shows that only very few cases are seriously documented and that most reports come just from hearsay. Fact is that very tiny (i.e., young) specimens of the fish *Vandellia cirrhosa* with a size of less than 2–4 cm in length may enter the genital openings of naked humans while swimming in fresh waters. Such penetrations, however, are also known from tiny leeches.

Keywords Candirus • Caneros • Ectoparasitism • Endoparasitism • *Vandellia cirrhosa* • Bloodsucking fish • Vampire fish • Body penetration • Catfish

What a horror story! When coming back from the regions of the Amazonas and its tributaries to their rather safe European countries, adventurers reported that they had been told by native people or by other roaming natural scientists that there exist bloodsucking fish (often swimming in swarms), which may enter the openings of the male and female genitalia of humans or are even able to penetrate into their anus. Due to several teeth at the mouthparts and due to backward pointing hooks at each cover of their gills, the penetrated fish are unable to slide backwards after having entered. Thus, they cannot leave from there, and while degrading they initiate severe infection leading to death, if they are not extracted by surgery. The native population living along the shores of the river Amazonas and Orinoco used in former times a composition of different plants to get rid of such invaded fish,

H. Mehlhorn (✉)
Institute for Parasitology, Heinrich Heine University, Universitätsstr. 1, 40225 Düsseldorf, Germany
e-mail: mehlhorn@uni-duesseldorf.de

S. Klimpel and H. Mehlhorn (eds.), *Bats (Chiroptera) as Vectors of Diseases and Parasites*, Parasitology Research Monographs 5, DOI 10.1007/978-3-642-39333-4_10,

however, the results were mostly poor and in many cases intoxication occurred due to poisonous ingredients in the plants. Such invasions apparently occurred when naked persons swam or stayed in waters while hunting larger fish for food (De Castelnau 1855; von Martius 1829 cited by Piper 2007; Bulanger 1898). However, clearly documented cases are rare (Gudger 1930; Herman 1973; Bréault 1991) and other reports probably deal with bloodsucking leeches, which rather often enter the nose, mouth, or the openings of the genitalia of swimming humans or of animals that drink water while standing at the shore.

At first, it was believed that these fish, that had entered naked humans, were attracted by human urine when it was excreted during swimming or while staying in fresh waters in South America, which are the biotopes of these fish (Herman 1973). Experiments of Spotte's group (Spotte et al. 2001), however, proved that the initial idea was wrong, since it was observed that these fish hunt just by sight looking for larger fish when lurking hidden in sand or in mud of shallow waters. Having reached a prey fish they enter the space below the cover of the gills and seem to suck blood there. This is the rule, while the horrorful imagination that fish may enter body openings of humans becomes reality surely only in very rare cases, when fish are doing it erroneously, since they are of course firmly adapted to their fish hosts already since millions of years, whereas humans are absolutely late comers on earth and in addition they are extremely rare in the floods of the Amazonas river and thus unable to serve as a regular prey. Since once entered fish will mostly die in the body hollow of men, there is also no chance that fish might learn to penetrate further humans.

But also the idea that these so-called vampire fish suck blood actively turned out to be wrong, when the anatomy of these worm-like appearing fish was studied. It was found that these fish, which belong to the group of catfish (Siluriformes) and reach a maximum length of about 15 cm, wait hidden in the sand or mud for larger fresh water fish and attach at the prey's aorta that leaves the heart in order to transport the blood first to the gills and later to the organs. By help of their needle-like teeth, the candiru fish perforates the blood vessel, which by its own inner pressure pumps blood into the stomach of the parasite, which then becomes filled up within a few seconds. After this short feeding period of 2 min at the maximum, the fish detaches from its host and hides again in the sand or mud, where digestion occurs "leisurely" until hunger comes again and another possible prey fish swims by. Inside human genitalia, the candiru fish, however, are unable to take up masses of blood, since they do not possess a sucking respectively pumping system in their mouth or in their anterior intestinal system.

These fish, which in their local regions at the Amazonas and Orinoco are called "candirus" respectively "caneros," are systematically classified by the famous French scientists Cuvier and Valenciennes, who noted them in 1846 in their famous fish system belonging to the family of catfish (Siluridae) as *Vandellia cirrhosa*. **Thus, the discovery of the candirús and their life cycle shows that very often animals may spread fear and horror until their true life cycle and their common behaviour is known**.

Simplified systematic position of *Vandellia cirrhosa*	
Cladus	Vertebrata
Classis	Osteichthyes (fish with bones)
Order	Siluriformes (catfish)
Family	Trichomycteridae
Subfamily	Vandelliinae
Genus	*Vandellia*
Species	*Vandellia cirrhosa*
First description:	
Cuvier and Valenciennes, 1846	
Local names:	
Candirù, Canero; ♀/♂ 15–17 cm long	

References

Bréault IL (1991) Candiru: Amazonian parasitic catfish. J Wilderness Med 2:304–312

Bulanger GA (1898) Exhibition of specimens and remarks upon the habits of the siluroid fish *Vandellia cirrhosa*. Proc Zool Soc Lond 3:90–103

De Castelnau F (1855) Expéditions dans les parties centriques de l'Amérique du Sud, 1843 à 1847: Animaux nouveaux ou rares. Zoologie 3:50–60

Gudger EW (1930) Bookshelf browsing on the alleged penetration of the human urethra by an Amazonian catfish called Candiru. Am J Surg 8:170–188 and 443–457

Herman JB (1973) Candiru: urinophilic catfish – its a gift to urology. Urology 1:265–267

Piper R (2007) Extraordinary animals: an encyclopedia of curious and unusual animals. Greenwood Press, London

Spotte S, Petry P, Jansen ZAS (2001) Experiments on the feeding behaviour of the haematophagous candiru. Environ Biol Fishes 60:459–464

Index

S. Klimpel and H. Mehlhorn (eds.), *Bats (Chiroptera) as Vectors of Diseases and Parasites*, Parasitology Research Monographs 5, DOI 10.1007/978-3-642-39333-4,

Printed by Printforce, the Netherlands